RECAPITALIZING THE NAVY

A Strategy for Managing the Infrastructure

Committee on Shore Installation Readiness and Management
Naval Studies Board
Commission on Physical Sciences, Mathematics, and Applications
National Research Council

NATIONAL ACADEMY PRESS
Washington, D.C.   1998

This work was performed under Department of the Navy Contract N00014-96-D-0169/0001 issued by the Office of Naval Research under contract authority NR 201-124. However, the content does not necessarily reflect the position or the policy of the Department of the Navy or the government, and no official endorsement should be inferred.

Copies available from:

Naval Studies Board
National Research Council
2101 Constitution Avenue, N.W.
Washington, D.C. 20418

Printed in the United States of America

COMMITTEE ON SHORE INSTALLATION READINESS AND MANAGEMENT

JOHN D. CHRISTIE, Logistics Management Institute, *Co-Chair*
JOHN F. EGAN, Nashua, New Hampshire, *Co-Chair*
JOHN W. ASHER III, Strategic Marketing and Analysis, Inc.
ALBERT J. BACIOCCO, JR., The Baciocco Group
LLOYD A. DUSCHA, Reston, Virginia
ELVIN R. HEIBERG III, Heiberg Associates
SAMUEL D. KLEINMAN, Center for Naval Analyses
GENIE McBURNETT, Falls Church, Virginia
NANCY Y. MOORE, RAND
GERALD NADLER, University of Southern California
SEAN O'KEEFE, Syracuse University
DAN R. OLSEN, JR., Carnegie Mellon University
MICHAEL W. O'NEILL, Deloitte & Touche, LLP
HERBERT RABIN, University of Maryland
JOSEPH B. REAGAN, Saratoga, California
JOHN M. STEWART, McKinsey & Company, Inc.
RAYMOND M. WALSH, Sonalysts, Inc.
MITZI M. WERTHEIM, Center for Naval Analyses

Naval Studies Board Liaison
Seymour J. Deitchman, Chevy Chase, Maryland

Navy Liaison Representatives
VADM W.J. Hancock, USN, Office of the Chief of Naval Operations, N4 (through July 31, 1998)
VADM James F. Amerault, USN, Office of the Chief of Naval Operations, N4 (as of August 3, 1998)
RADM John T. Scudi, USN, Office of the Chief of Naval Operations, N46 (through August 17, 1998)
RADM Annette E. Brown, USN, Office of the Chief of Naval Operations, N46 (as of October 26, 1998)
David M. Wennergren, Office of the Chief of Naval Operations, N464 (through October 26, 1998)

Consultants
Sidney G. Reed, Jr.
James G. Wilson

Staff
Charles F. Draper, Program Officer

Marine Corps Liaison Representative
LtGen John E. Rhodes, USMC, Commanding General, Marine Corps Combat Development Command

Ronald D. Taylor, Director
Charles F. Draper, Program Officer
Susan G. Campbell, Administrative Assistant
Mary G. Gordon, Information Officer
Larissa M. Markarian, Senior Project Assistant (through October 16, 1998)

COMMISSION ON PHYSICAL SCIENCES, MATHEMATICS, AND APPLICATIONS

Preface

As the Department of the Navy plans to meet the challenges of the 21st century, it must take into account budget trends since the end of the Cold War and the need for maintenance and modernization of the fleet. To adjust, significant restructuring of both fleet and shore activities has been undertaken and will continue. However, such restructuring must be done without adversely affecting the ability of naval forces to execute their missions. A serious challenge to the Department of the Navy is how to recapitalize and modernize for the future while maintaining fleet readiness within projected budgets. Reducing the costs associated with maintaining an extensive shore establishment has been viewed by the Department of Defense and the Department of the Navy as one means for achieving the necessary cost savings to finance the fleet of the future.

Naval installations are major components of the shore establishment and are complex enterprises. Some are comparable to cities, with airports and harbors; others incorporate shipyards and aviation depots. Most have family housing, hospitals, and child care and commissary facilities. Typically, a base commander and his or her staff are responsible for managing more than 100 different activities and often must provide such support to numerous tenant organizations. Unfortunately, these commanders are not provided the tools needed for managing such complex enterprises, and today's accounting, information management, and personnel and legal systems are ill-suited to the challenge. By contrast, large enterprises in the private sector rely on management techniques and business practices that are based largely on advances in information technology, systems and industrial engineering, operations research, organizational design, accounting, production scheduling and economics, management of human resources, and environmental management. Use of these techniques has dramatically reduced overall operating costs and enabled better use of resources in major functions. Their application to shore installation operations could provide the same benefit to the Department of the Navy. With this in mind, the Navy has in fact established the Smart Base project, a set of initiatives to apply state-of-the-market, commercially available technology, policy changes, and better business practices to shore installation operations in an effort to increase efficiency. This initiative along with others is providing a testbed for new ways of doing business.

At the request of Admiral Jay L. Johnson, USN, Chief of Naval Operations (CNO) (see Appendix A), the National Research Council (NRC) conducted a study designed to assist the Department of the Navy with its ongoing efforts to improve shore installation operations, readiness, and management through the focused application and integration of state-of-the-market technologies and business methods (including outsourcing, privatization, and partnerships with state and local governments), with a goal of reduced cost of infrastructure. The Committee on Shore Installation Readiness and Management, operating under the auspices of the NRC's Naval Studies Board, was appointed to (1) identify business practices (or enterprise processes) in addition to application of technology for enhancing efficiency; (2) recommend how implementation might be accomplished and evaluate efficiencies that might be gained; and (3) provide estimates that project Navy-wide savings that could result from further application. Against these objectives, it also was requested that the committee examine the Navy's Smart Base project.

In responding to the CNO's request, the committee focused its considerations initially on U.S. Navy efforts to reduce shore installation costs, exclusive of base realignment and closure (BRAC). The committee's interpretation of the terms of reference was that it should investigate what could be done to achieve infrastructure savings outside of BRAC. The committee soon realized, however, that reengineering naval installations could bring only small savings and that the Navy and the Marine Corps infrastructure in total should be examined fully if the Department of the Navy's recapitalization goals are to be met. Furthermore, in its efforts to provide implementable recommendations, the committee identified and presents in the body of the report a number of specific actions that it believes are best assigned to particular individuals under the current Navy organization, e.g., the Secretary of the Navy, the Chief of Naval Operations, the Chief Information Officer, and such. In many cases, alternative approaches might be possible, particularly under a different organizational structure. The committee had no desire to comment on how the Navy is organized. Given the existing organizational structure, however, the committee felt compelled in many instances to provide at least one method by which specific problems could be solved and progress made in these complex areas.

The Executive Summary gives the report's major recommendations. The body of the report presents and discusses additional, specific, detailed recommendations and actions regarding naval installations and also key aspects of the multifaceted Navy infrastructure. In making its recommendations, the committee was very conscious of the impact that potential changes in the infrastructure might have on the ability of the operating forces to carry out their missions. The committee believes that making the infrastructure more efficient by using good business practices, as opposed to the current approach of arbitrarily reducing funding for the infrastructure, will in fact enhance the support of the operating forces and improve the capability of the infrastructure to respond to new and/or additional requirements. This would include the requirement for dealing with sudden emergencies that necessitate rapid response, such as Desert Storm.

The committee first convened early in 1998 and met for approximately 8 months. During that time, it held the following meetings and visited the following bases:

- February 4-5, 1998, in Washington, D.C. Organizational meeting. Navy briefings.
- March 4-5, 1998, in Washington, D.C. Navy briefings. Managerial accounting and facility management in the commercial sector.
- April 27-30, 1998, in San Diego, California. Site visit to assess the regionalization efforts underway at Naval Base, San Diego. Briefings on management and change.
- May 20-21, 1998, in Washington, D.C. Navy briefings. Use of information technology in the U.S. Air Force.
- June 4, 1998, in Pascagoula, Mississippi. Subcommittee site visit to Navy's Smart Base.
- June 10-11, 1998, in Washington, D.C. Navy briefings. Briefings on management and change.
- June 24, 1998, in Portsmouth, New Hampshire. Subcommittee site visit to Navy's Smart Industrial Base.
- July 1-2, 1998, in Washington, D.C.
- August 11-12, 1998, in Washington, D.C.

The resulting report represents the committee's consensus view on the issues posed in the charge.

Acknowledgments

The Committee on Shore Installation Readiness and Management extends its gratitude to the many individuals who provided valuable information and support during the course of this study. Special acknowledgment goes to Mr. David M. Wennergren, who assisted the committee with countless briefings and information throughout the early stages of this study.

The committee wishes to extend a special thanks to RADM Veronica Z. Froman, USN, Commander, Naval Base, San Diego. Admiral Froman and her staff were gracious in hosting the committee on its 4-day site visit to learn more about regionalization efforts in the Southwest Region. Likewise, the committee wishes to thank VADM Henry C. Giffin, USN, Commander, Naval Surface Force, U.S. Atlantic Fleet, and RADM R. Tim Zeimer, USN, Commander, Naval Base, Norfolk, for visiting with the committee in Washington, D.C., to discuss fleet readiness and regionalization efforts in the Hampton Roads region.

The committee also wishes to thank CAPT Vernon T. Williams, USN, Shipyard Commander, Portsmouth Naval Shipyard, and CDR Christy J. Wheeler, USN, Commander, Naval Station, Pascagoula, for hosting site visits by members of the committee to learn more about initiatives surrounding the Navy's Smart Base project.

Finally, the committee wishes to thank the many men and women throughout the Armed Services, as well as government, academic, and industry leaders who provided the committee with insightful discussions throughout the course of this study. Without their combined efforts, the committee's report would not have been possible.

Acknowledgment of Reviewers

This report has been reviewed by individuals chosen for their diverse perspectives and technical expertise, in accordance with procedures approved by the National Research Council's (NRC's) Report Review Committee. The purpose of this independent review is to provide candid and critical comments that will assist the authors and the NRC in making the published report as sound as possible and to ensure that the report meets institutional standards for objectivity, evidence, and responsiveness to the study charge. The content of the review comments and the draft manuscript remain confidential to protect the integrity of the deliberative process. We wish to thank the following individuals for their participation in the review of this report:

ADM Stanley R. Arthur, USN (retired), Lockheed Martin Corporation,
MG Norman G. Delbridge, USA (retired), Springfield, Virginia,
Brian K. Dickson, PricewaterhouseCoopers LLP,
L. Paul Dube, Arlington, Virginia,
John R. Kreick, Sanders, a Lockheed Martin Company (retired),
Joe H. Mize, Oklahoma State University,
Richard L. Tucker, University of Texas, Austin, and
Anthony M. Valletta, SRA International, Inc.

Although the individuals listed above provided many constructive comments and suggestions, responsibility for the final content of this report rests solely with the authoring committee and the NRC.

Contents

Executive Summary

The Department of the Navy wants to improve shore installation operations, readiness, and management by skillfully leveraging state-of-the-market technologies and business methods such as outsourcing, privatization, and partnerships with state and local governments, with a goal of reduced cost of infrastructure. For the Navy itself, where all forces float[1] or fly, the shore establishment is synonymous with infrastructure, which includes "all activities that provide support or control of forces from fixed bases of operation."[2]

In response to a Navy request, the Committee on Shore Installation Readiness and Management, operating under the auspices of the Naval Studies Board, was created to offer advice on how to accomplish the goal of providing quality infrastructure support at significantly less overall cost to the fleet. The committee was asked to (1) identify business practices (or enterprise processes) and technology applications that could materially enhance the efficiency and effectiveness of operations; (2) recommend a plan for implementing the changes needed and provide approximate estimates of the efficiencies that might be gained from implementing that plan; and (3) develop estimates of the potential Navy-wide savings that could result from extending the committee's recommendations to other areas. Against these objectives, it also was requested that the committee examine the Navy's Smart Base project.

---

[1]The term "float" includes submarines.

[2]Graves, T.J., D. Drake, P. Forsyth, and J.L Wilson. 1995. *A Reference Manual for Defense Mission Categories, Infrastructure Categories, and Program Elements*, Paper P-3133, Institute for Defense Analyses, Alexandria, Va., June.

THE CHALLENGE FACING THE U.S. NAVY

To accommodate lower post-Cold War budget levels, the Navy significantly reduced its modernization funding. The Navy leadership now needs about an additional $3.5 billion to $5.0 billion per year to recapitalize and modernize for the future while maintaining fleet readiness, and the Department of Defense (DOD) has identified infrastructure savings as the source for funding this shortfall.[3] If the Navy is to maintain its current fleet size and recent levels of peacetime deployment and peacekeeping operational activity with constant or declining budgets, the only available source of funds for modernization is the infrastructure.[4] Currently, infrastructure activities account for $28 billion (or 40 percent) of the annual Navy budget of about $70 billion. This is an increase from about 37 percent in FY 1991.

THE NEED FOR A CORPORATE STRATEGY AND LEADERSHIP

Past performance and the committee's review of ongoing initiatives indicate that reallocating $3.5 billion to $5.0 billion annually from Navy infrastructure activities for recapitalization and modernization will require a more extensive effort than is currently underway. Current initiatives such as regional consolidation of installation management functions, designation of regional maintenance coordinators, and the Smart Base project are a good start, but they will not provide the desired savings. There are two primary reasons for this conclusion:

- First, ongoing initiatives are focused on only about one-third of the infrastructure and are projected to result in a maximum of about $750 million in annual infrastructure cost reductions—well short of the $3.5 billion to $5.0 billion annual goal. Thus, there is little likelihood that today's initiatives could by themselves solve the problem at hand.
- Second, and perhaps more important, is the fact that the committee could not identify an overall corporate Navy strategy for solving the problem. The important changes that are underway are being led by individual staff activities and support elements that lack the authority to change the requirements for or the methods of providing goods and services to the fleet. Moreover, many of the reductions made thus far appear to be pro rata cuts rather than being based on solid research, analyses, and assessments of risk.

---

[3]U.S. General Accounting Office. 1997. *Defense Infrastructure,* GAO/HR-97-7, Washington, D.C., February.

[4]If defense and Navy budget levels were not determined in large part by domestic and political considerations, the linkage between modernization needs and infrastructure efficiencies could be broken, but that is unlikely in the current national security environment.

Major Recommendation: *To achieve its recapitalization funding goal, the Navy should develop and implement a corporate-wide strategy to improve the business operations of the entire Navy infrastructure. The senior leadership of the Navy, led by the Secretary of the Navy and the Chief of Naval Operations (CNO), should establish a clear vision and a corporate-wide strategy for conducting the future operations of the entire naval system within the budget constraints projected. The strategy must be clear on what is to be achieved, in concrete terms, how it is to be achieved, with what means it is to be achieved, and when it is to be achieved. The strategy, of necessity, must address all portions of the Navy infrastructure, not just a few isolated portions thereof. A compelling case for major change in the way business is conducted must be made by the CNO and communicated to all elements of the Navy. Responsibilities and authorities to implement change must be made clear and issued by the CNO.*

Experience in both the public and private sector shows that such "enterprise reengineering" cannot be implemented easily. The required efforts go far beyond the authority of individual staff elements in the Office of Naval Operations (OPNAV) or the Secretariat (e.g., Shore Installation Management Division (OPNAV-N46), Deputy Chief of Naval Operations for Logistics (N4), Assistant Secretary of the Navy, Financial Management and Comptroller (ASN/FM)), as the committee understands these activities today.

Additional insights and recommendations on what is needed to achieve change across the entire Navy system are provided in the body of this report. In developing its corporate-wide strategy the Navy should aggressively reexamine long-standing business practices that are deeply ingrained in its culture. The committee believes that, to provide visibility, foster innovative solutions to today's complex problems, and measure progress, the Navy's efforts should focus on implementing three key interrelated enablers.

NAVY-WIDE ENABLERS

Three enablers—performance metrics, information systems technology, and competition—are singularly important to implementing a cohesive strategy and plan that will achieve the degree of change required in this instance.

Performance Measures, Cost Management, and Allocation of Resources

Cost visibility and performance measurement to gauge progress toward meeting mission goals are critical to good decision making and allocation of resources. Traditionally, DOD and Navy resource allocation processes have, for the most part, focused on input cost figures that amount to planning factors. With regard to infrastructure, for example, the requirement for funding real property maintenance is often expressed as a percentage of the current plant value (e.g., real property maintenance "should be funded" at 2 percent of the current plant value). Such requirements do not result from analysis of actual needs; they do not identify the consequences of funding at higher or lower levels, nor do they describe the consequences or loss in capability associated with not funding to meet the "requirement."

To improve management of the infrastructure, particularly from a business perspective, the Navy will have to develop and use performance measures that are important to the operational forces. Such measures must facilitate assessments of infrastructure support outputs as they affect force readiness to carry out fleet missions. Output measures are also essential to making more informed judgments regarding alternative infrastructure investments and the consequences of those alternatives. To facilitate implementation, the committee provides a suggested seven-step process for developing performance measures that involve both service providers and service users.[5] The process will allow decision makers, from the program manager to the Chief of Naval Operations, to evaluate what is required to support the user and what level of performance should be expected by the user.

The Navy Working Capital Fund (NWCF), which currently provides more than $20 billion in goods and services annually to Department of the Navy customers, lacks cost visibility and output measures for assessing performance. NWCF operations are overseen by financial managers rather than line managers, and there is no single individual or chief operating officer who is dedicated to overseeing and improving management of the total system.

Major Recommendation: *The Navy should establish a management information system to track support-system performance and costs. This system should be based on an integrated set of cost and performance metrics that are developed using fleet user inputs as well as those of service providers.*

More detailed information and recommendations related to the components of a management information system are included in Chapter 3 of this report.

In addition, accurate vision statements are important when formulating long-range plans and measuring progress. The committee believes that the motto in the current *21st Century Shore Support Infrastructure: Navy Infrastructure Vision and Strategic Plan,*[6] "Equal to or better service at equal to or less cost," is inconsistent with the goal of reducing the infrastructure to the minimum essential necessary to meet users' requirements.

---

[5]See Appendix C and discussion in Chapter 3.

[6]Hancock, VADM W.J., USN, Deputy Chief of Naval Operations for Logistics (N4). 1997. *21st Century Shore Support Infrastructure: Navy Infrastructure Vision and Strategic Plan,* Washington, D.C., June 14.

Major Recommendation: *The Navy should change its statement of infrastructure vision to "Essential service at minimum cost."*

Information Technology for Infrastructure Management

Information technology, when used creatively, can enable organizations to provide services in ways not previously possible. It can also assist managers to better align lower-level responsibilities and authorities, and thus can save money by eliminating work. Numerous private-sector firms have significantly improved their global competitiveness by reengineering their business processes and management structures. Improved competitiveness in this regard has typically involved adopting a customer-oriented focus and skillfully leveraging information management and communications in order to reduce the overall costs of operations. To succeed, this approach requires an enterprise-wide perspective of "what is" and "what needs to be." A clear vision of the desired management structure, and how it needs to function, is central to any such reengineering effort.

To improve the efficiency and effectiveness of its infrastructure operations, the Navy must fully identify both the individuals and the organizations that will be responsible and accountable for specific infrastructure outputs and the content of the associated management information. A substantial effort must be made to develop and maintain networked information systems, including World Wide Web-based tools, databases, and applications that will enable integrated management of the Navy infrastructure. The basic task is to provide all essential users in the infrastructure and the fleet with desk-to-desk connectivity and sufficient bandwidth such that they will be able to share data and services as they can now share voice communication.

Connectivity and access to shared information increase the value of services to users. With total connectivity, information technology has the potential for integrating ship and shore operations and enabling the infrastructure to supply services in ways not possible before. Information posted on the World Wide Web could replace most hard-copy manuals and general information materials. Providing services via a Navy-wide information space (infospace) rather than physically co-locating services and customers can effectively move the services from "down the street" to "on the desk." The concept of a Navy-wide infospace is that of an adaptive system defined at any time by a set of performance standards for timely and effective information delivery throughout the Department of the Navy. Used well, information technology saves money by eliminating work, not by moving it or automating it. Information technology can also help to conserve physical resources by providing ways to model the consequences of alternative decisions. A simple example is the electronic tool for ship berthing and movement developed and used at Naval Base, San Diego. It is cheaper to move ships on screens than under real power.

The Navy has spent billions of dollars on obtaining connectivity, but the primary focus is on the operational users, while the needs of infrastructure users are not being fully addressed. The result is that the benefits of full Navy-wide connectivity are not being realized even though the major portion of the costs are being incurred.

Major Recommendation: *The Navy should define and implement the concept of a Department of the Navy-wide information space (infospace) using a set of standards—some of which are now being partially implemented by the Information Technology for the 21st Century (IT-21) initiative—to serve both shore and afloat activities. The infospace should be defined by performance standards for timely and effective delivery of information and services throughout the Department of the Navy, including the fleet and its support. It should include a network infrastructure and a standard information workstation for every responsible information worker in the Navy, not just those in the operating forces. The infospace should be used as the primary vehicle for delivery and integration of information about the entire Navy infrastructure to include the following functions: acquisition, central logistics, central personnel, communications, force management, installations, medical functions, quality of life, science and technology, and training. Funding for this effort, including funding for essential technology upgrades and related training, should be identified and protected within the Department of the Navy.*

More detailed information and recommendations related to the components of a future Navy-wide infospace, with specific discussion of its importance for reducing infrastructure costs, are included in Chapter 3 of this report.

Competition

Competition can reduce the costs of providing services. In addition to achieving greater efficiencies, there are numerous advantages to competition.[7] From 1978 to 1994, the DOD held more than 2,000 public-private competitions that resulted in an average savings of 30 percent, or about $1.5 billion annually.[8] Within the Department of the Navy, competitive sourcing (i.e., competitive bidding among service providers) following Office of Management and Budget Circular A-76 procedures was used to conduct large numbers of detailed bottom-

---

[7](1) Competition forces public (or private) monopolies to respond to the needs of their customers; (2) competition rewards innovation—monopoly stifles it; (3) competition boosts the pride and morale of public employees (Osborne, David, and Ted Gaebler. 1992. *Reinventing Government,* Addison-Wesley Longman, Inc., Reading, Mass., pp. 80-84).

[8]Trunkey, R. Derek, Benjamin P. Scafidi, Francis P. Clark, Cheryl Kandaras, Andrew M. Seamans, LCDR Carolyn M. Kresek, USN, Robert P. Trost, Angela L. King, Christine H. Baxter, Kerensa E. Riordan, Steven Smith, and Michael Ye. 1998. *Moving Forward with A-76 in the Navy,* Research Memorandum 98-9.10, Center for Naval Analyses, Alexandria, Va., May.

up reviews, with each competition involving small numbers of positions. The A-76 procedures, in and of themselves, discouraged managers from taking action. In addition to this constraint, competition for many infrastructure functions was also limited by Navy policies, particularly those regarding the definition of positions as inherently governmental and those reserved for sea-shore rotation. The end result of these constraints and policies is that competitive sourcing, as it has been implemented, is not likely to be a major contributor to reducing infrastructure costs. This is unfortunate because the committee's review clearly suggests that the Navy could reduce its future infrastructure costs considerably by adopting a top-down rather than a bottom-up approach, and by readdressing long-standing management practices and policies with regard to an objective definition of billets suitable for competition (civilian and military), given the costs of infrastructure personnel.

The rotation of the career enlisted force between sea and shore billets is an integral part of the Navy's overall personnel management system. The sea-shore rotation policy has fundamentally been implemented by reserving many shore jobs for military personnel. In recent years, however, sea-shore rotation objectives increasingly have come into conflict with Navy objectives to reduce the number of support personnel ashore.

The metric used by the Navy to monitor sea-shore rotation is the sea-shore ratio. This ratio is based primarily on considerations of morale and retention, with planning factors based on "past experience" and conventional wisdom—rather than any direct cause-and-effect relationships—used as a guide. The ratio is not based on the kinds of analyses that major modern enterprises use to maximize the efficiency and effectiveness of their personnel management.

Major Recommendation: *The Navy should use competitive sourcing as a preferred approach to selecting the best providers of all support. In this regard, the Navy should establish a cross-functional team under the Assistant Secretary of the Navy/Installation and Environment (ASN/I&E) and the Assistant Secretary of the Navy/Research, Development, and Acquisition (ASN/RDA) to be responsible for overseeing the execution of competitive sourcing in business operation areas approved by Navy leadership.*

In addition, the Navy should address all existing constraints on sea-shore rotation. The CNO should broaden the objective to managing seagoing personnel as a part of total naval personnel management and should direct relevant elements of the Office of Naval Operations (OPNAV) and second-echelon commands to search for innovative ways to satisfy the morale and retention needs that allow greater flexibility in reducing the cost of the infrastructure.

More detailed information and recommendations related to competition are included in Chapter 3 of this report.

CLOSING COMMENT

In seeking to prepare now for the future, the Navy must assume that its budgets will at best remain nearly flat in constant-dollar terms. Thus, in order to free resources to fund essential modernization requirements, the Navy should aggressively seek to significantly reduce the cost of infrastructure operations, because closing additional bases is an option that the Congress will not support at this juncture. Failure to act quickly and comprehensively will most likely result in an inability to acquire sufficient combat platforms, weapons, and supporting systems to maintain the current force structure and keep the Navy preeminent in the future.

1

Introduction

BACKGROUND

Current U.S. defense strategy[1] calls for continuing to shape the strategic environment to advance national interests, maintaining the capability to respond to the full spectrum of anticipated current threats, and preparing for the threats of tomorrow. Implementation depends on the fundamentals of military power: quality people, ready forces, and superior organization, doctrine, and technology. The challenge is to construct an effective defense establishment with limited financial resources in accordance with Department of Defense guidance.[2,3] The U.S. Navy is meeting this challenge in part through the use of better business practices and the application of advanced technologies within its shore establishment. Pressures to do so come from competition between the need to maintain force readiness for the present and the need to modernize systems and technology for the fleet of tomorrow.

For the U.S. Navy, the shore establishment includes the fixed activities of the Navy first- and second-echelon commands with their subordinate units

---

[1]The White House. 1997. *A National Security Strategy for a New Century,* U.S. Government Printing Office, Washington, D.C., May. Available online at <http://www.whitehouse.gov/wh/eop.nsc/strategy>.

[2]Cohen, William S., Secretary of Defense. 1997. *Report of the Quadrennial Defense Review,* Office of the Secretary of Defense, Washington, D.C., May. Available online at <HtmlResAnchor http://www.defenselink.mil/pubs/qdr/>.

[3]Cohen, William S., Secretary of Defense. 1997. *Defense Reform Initiative Report,* Office of the Secretary of Defense, Washington, D.C., November. Available online at <http://www.defenselink.mil/pubs/dodreform/>.

located throughout the world. Naval installations are major components of the shore establishment and are complex enterprises. Some are comparable to cities, which perform many different functions while supporting numerous tenant organizations and often include child care and commissary facilities, family housing, hospitals, and so on.

The Office of the Chief of Naval Operations (CNO) Shore Installation Management Division (OPNAV-N46) is the CNO's lead office for Navy shore installation programs.[4] Its primary responsibilities include installation operations; base operating support; quality-of-life mission support; infrastructure vision, strategic planning, plans, and policy; housing; and the Navy's Smart Base project. The Deputy Chief of Naval Operations for Logistics (N4) is leading a campaign to reduce the cost of the infrastructure through an improvement in efficiency and the consolidation of activities encompassing naval installations.[5] This campaign includes initiatives such as regionalization (the consolidation of base operating support functions in regions where individual installations and facilities formerly operated more independently) and Smart Base (a collection of experiments and tests designed to reduce costs and improve the delivery of support services at naval shore installations). Although some of these initiatives are projected to result in significant savings, the committee estimates that these initiatives will most likely yield annual savings of no more than about $500 million. This shortfall is due primarily to the fact that the managers of these naval installations (i.e., base commanders and regional commanders) control only a small fraction of all the resources located within their installations. For example, in the San Diego region, with about 56,000 military and civilian Navy employees and an annual personnel cost of about $2.5 billion, the regional base commander has control of less than $500 million of Navy costs per year. The vast majority of annual costs at naval installations are controlled by other major commands that may or may not be tenants of the specific installation. These "claimants" are the Navy organizations that are responsible for expending the portions of the Navy budget apportioned to them. Examples of claimants are the Pacific Fleet (PACFLT), the Naval Sea Systems Command (NAVSEA), and the Commander, Navy Education and Training (CNET).

Partitioning of the responsibility for financial resource management is the reason that initiatives limited to base commanders, and to the supporting functions under their control, are estimated to result, at best, in limited ($500 million) annual infrastructure cost reductions. The relatively small potential cost reduc-

---

[4]Deputy Chief of Naval Operations, Logistics (N4) and the Office of the Deputy Chief of Staff for Installations and Logistics. 1997. *Navy and Marine Corps Annual Logistics Review,* Washington, D.C., p. 9.

[5]Hancock, VADM W.J., USN, Deputy Chief of Naval Operations for Logistics (N4). 1997. *21st Century Shore Support Infrastructure: Navy Infrastructure Vision and Strategic Plan,* Washington, D.C., June 24.

tions from shore installation management initiatives compared to the overall size of the Navy infrastructure ($26.7 billion in FY 1999) motivated the committee to broaden its perspective on the problem at hand.

With respect to the U.S. Navy, where all forces float[6] or fly, the shore establishment in the broadest sense is synonymous with infrastructure, which—according to the Joint Chiefs of Staff—includes "all activities that provide support or control of forces from fixed bases of operations."[7] In accordance with this broader perspective of the Navy shore establishment, the committee assesses below the size of the basic problem driving the Navy's desire to reduce infrastructure costs and reexamines the potential base from which infrastructure cost reductions could be obtained. This latter examination places the committee's review of N4 related activities in perspective. It also includes a discussion of recent trends in infrastructure costs and the resources associated with individual infrastructure functions.

THE NAVY'S PROBLEM: MEETING ITS RECAPITALIZATION AND MODERNIZATION GOAL THROUGH INFRASTRUCTURE REDUCTIONS

Since the end of the Cold War, the Department of the Navy, along with the other Services, has been faced with addressing new national security challenges with smaller budgets. As the Department of the Navy reduced its overall force structure, it cut modernization funding by a larger percentage than other categories. Infrastructure costs have been cut, but there is general agreement among Navy leadership that infrastructure can and should be reduced more, compared to the fleet that it supports. In this sense, further infrastructure cost reductions are a potential source of recapitalization funding to sustain the desired fleet capability.

From FY 1991 to FY 1999 the Department of the Navy budget has been reduced by approximately $40 billion (or about 33 percent), and current projections indicate that it will continue to decline at a rate of about 1.4 percent per year.[8] Figure 1.1 illustrates the trends in funding for U.S. Navy programs since FY 1991. Most notably, force modernization was reduced by approximately $16 billion (about 42 percent) through FY 1999. This 42 percent reduction in modernization compared to a 33 percent department-wide reduction in total funding amounts to about a $3.5 billion excess reduction in the U.S. Navy's moderniza-

[6]The term "float" includes submarines.

[7]Graves, T.J., D. Drake, P. Forsyth, and J.L. Wilson. 1995. *A Reference Manual for Defense Mission Categories, Infrastructure Categories, and Program Elements*, Paper P-3133, Institute for Defense Analyses, Alexandria, Va., June.

[8]All data, findings, and figures shown in this chapter are a result of the committee's use of FY 1999-2003: President's Budget Future Years Defense Programs. See Appendix B for further details.

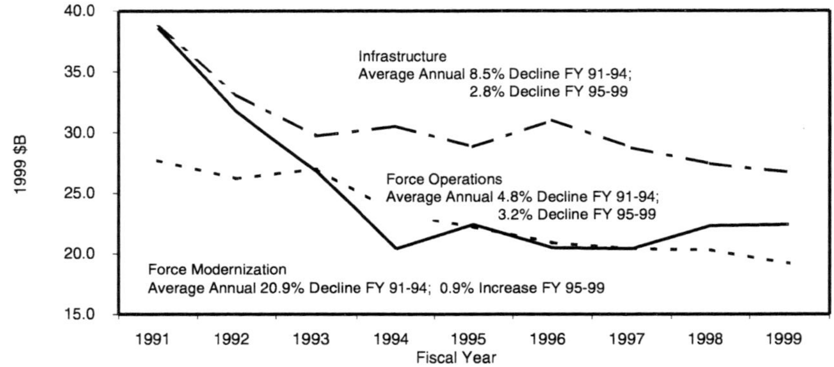

Fiscal Year	1991	1992	1993	1994	1995	1996	1997	1998	1999
Modernization	38.6	31.7	26.8	20.4	22.4	20.5	20.4	22.3	22.4
Operations	27.7	26.2	27.0	23.3	22.2	20.9	20.4	20.3	19.2
Infrastructure	38.7	33.1	29.7	30.5	28.8	31.0	28.7	27.4	26.7
Total	105.0	91.0	83.5	74.2	73.4	72.4	69.5	70.0	68.3

FIGURE 1.1 Funding for U.S. Navy programs, 1991-1999.

tion accounts. To return the Navy modernization budget to its former (FY 1991) share of the total Department of the Navy budget would require an estimated additional $3.5 billion per year (above the FY 1999 figure of $22.4 billion). It is the committee's judgment that establishing a reinvestment goal of $3.5 billion to $5.0 billion is a reasonable target. Recent analysis of defense modernization needs for ship construction through 2001 cited a range of approximately $4.0 billion to $5.0 billion per year (more than is currently projected) for modernization.[9] Clearly, if the Department of the Navy is to achieve its future recapitalization and modernization objectives, a reduction in infrastructure and associated costs will be necessary. If the $3.5 billion to $5.0 billion annual shortfall for Navy modernization is to be recovered entirely from the Navy infrastructure, a 13 to 19 percent reduction in infrastructure costs would be needed. Reductions of this magnitude will demand significant change in the overall management of the infrastructure within the Department of the Navy. As defined by the Office of the Secretary of Defense (OSD), infrastructure comprises the following functional categories: acquisition, central logistics, central personnel, communications, force management, installations, medical functions, quality of life, science

[9]Lopez, VADM Thomas J., USN, Deputy Chief of Naval Operations, Resources Warefare Requirements and Assessments (N8). 1995. "The State of the Navy," briefing to the Naval Studies Board, September 13.

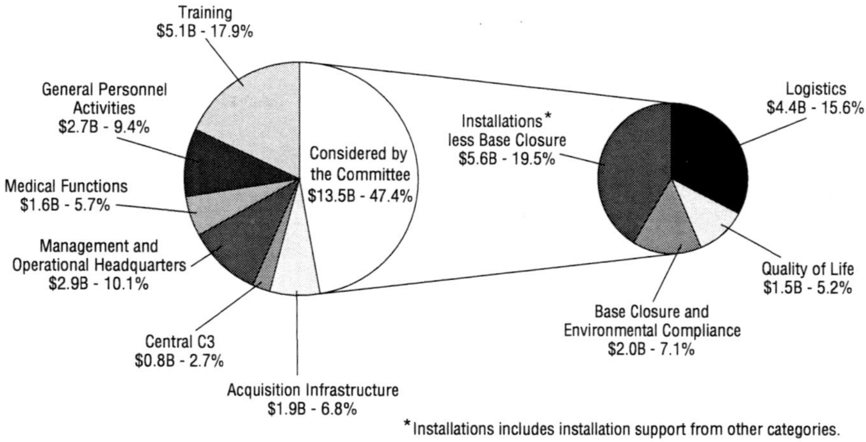

FIGURE 1.2 Navy infrastructure initially considered by the committee. Data shown are average values from FY 1995 to FY 1999.

and technology, and training.[10] Because the N4 is one of the more proactive staff elements sponsoring programs specifically with the aim of infrastructure cost reduction, the committee focused its attention on those functional categories most closely associated with the N4, namely, installations (less base closure), central logistics, quality of life (which is largely family housing), and base closure and environmental compliance. Figure 1.2 illustrates those infrastructure categories initially considered by the committee.

Trends in infrastructure categories initially reviewed by the committee are illustrated in Figure 1.3. Base closure and environmental compliance grew significantly from FY 1991 to FY 1995 and then declined substantially from FY 1996 through FY 1999. The Navy does not have much influence on funding for these programs. Conversely, installations (excluding base closure) make up almost 20 percent of the infrastructure and are the subject of the Navy's regionalization initiative. As indicated above, the committee estimates that this concept could eventually yield savings on the order of 10 percent, or $500 million per year, but realization of these savings will demand the full participation of all commands and strong support from the CNO. To yield such significant savings, reengineering of the services provided by the installations is a necessary next step. Logistics comprises mainly central supply, transportation, and maintenance activities. Each of these areas has initiatives aimed at streamlining services,

[10]Office of the Secretary of Defense, Program Analysis and Evaluation. 1996. Defense Infrastructure Overview Briefing, November 1.

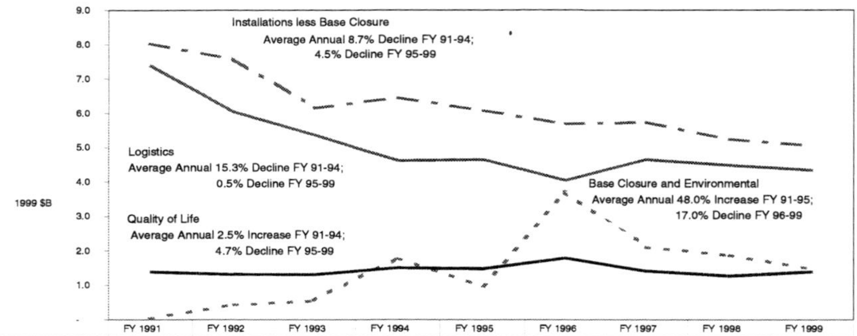

	FY 1991	FY 1992	FY 1993	FY 1994	FY 1995	FY 1996	FY 1997	FY 1998	FY 1999
Base Closure	0.1	0.4	0.6	1.8	1.0	3.7	2.1	1.9	1.5
Installations	8.1	7.6	6.2	6.5	6.1	5.7	5.7	5.2	5.1
Logistics	7.4	6.0	5.4	4.6	4.7	4.0	4.7	4.5	4.3
Quality of Life	1.4	1.3	1.3	1.5	1.5	1.8	1.4	1.3	1.4
Total	16.9	15.4	13.4	14.4	13.2	15.2	13.9	12.9	12.3

FIGURE 1.3 Trends in funding for areas of infrastructure considered by the committee.

but much more is needed. For example, in the area of maintenance, regional maintenance coordinators for the most part have only the power of personal persuasion to make changes and achieve savings. Family housing makes up about two-thirds of the funding for the functional category "quality of life." Public-private ventures can have a large impact on the cost of meeting this requirement and should be pursued as a matter of priority.

The recent actual and projected total Navy infrastructure decline of about $750 million per year (shown in Figure 1.1) is primarily a result of money being arbitrarily taken out of the budget, particularly in base operating support, rather than the result of the integrated introduction of new innovations to reduce the cost of infrastructure. In fact, budget pressures have been the driving impetus in forcing innovation and overcoming resistance to change. Although a forcing function is desirable, it is also necessary to have an overall, integrated strategic plan, which appears to be lacking, that includes metrics that measure the output of the infrastructure. Without such a plan and metrics, it will not be possible to determine where additional savings can be obtained with acceptable risk or where potential budget reductions would create unacceptable risks and reductions in fleet capabilities and readiness.

The $750 million in annual infrastructure reductions projected over the past few years by the U.S. Navy is a long way from the $3.5 billion to $5.0 billion annual goal. Meeting such a goal will necessitate addressing all functional categories of the infrastructure, which in turn requires a strategy for managing the overall infrastructure.

Although training and general personnel activities were not initially reviewed by the committee, these functional categories do account for about one-fourth of the total cost of Navy infrastructure. They also influence every other category of the infrastructure and are worthy of a separate review and/or study.

The committee was briefed on U.S. Navy training activities after making a specific request to obtain information on training-related initiatives designed to reduce infrastructure costs, but it failed to find any cost reduction initiatives comparable to those being developed and supported in the installation management area. Personnel management is another area that was not considered in detail; however, it is apparent that sea-shore rotation objectives have come into conflict with U.S. Navy objectives to reduce the number of support personnel ashore and thereby reduce infrastructure costs (see Chapter 3).

ORGANIZATION OF THIS REPORT

A review and assessment of selected U.S. Navy initiatives to reduce infrastructure costs are presented in Chapter 2. As indicated above, the review focuses on the initiatives being supported within the OPNAV-N4 establishment for two reasons: (1) the leadership within the functional portions of the OPNAV staff for reducing infrastructure costs lies in OPNAV-N4, and (2) the committee's attempts to obtain information on significant infrastructure cost reduction initiatives in other functional categories did not bear fruit.

Chapter 3 points to the need for an overall strategy for managing the Navy's infrastructure. As paraphrased from the words of General Andrew Goodpaster, a strategy contains three elements: (1) What is to be achieved? (2) How is it to be achieved? and (3) With what will it be achieved? In this case the "what" is to reduce the Navy infrastructure in order to recapitalize and modernize the Navy. The "how" is by reengineering the business practices of the infrastructure and applying technology to the infrastructure to enhance efficiency. The "with what" is, in particular, what the committee found lacking as it reviewed individual initiatives. Specifically, the integrating threads—a comprehensive strategy—to tie disparate, and sometimes overlapping, initiatives and activities together were lacking.

Chapters 3 and 4 indicate that two objectives are key to the Navy's success in meeting its recapitalization and modernization goal: (1) assessing and specifying the key enablers that the Navy should use to implement its chosen strategy (i.e., specifying the third, "with what," portion of a coherent strategy) and (2) defining top leadership's role and potential actions needed to achieve a major change of the magnitude required.

Chapter 3 covers a strategy for implementing three key enablers that are essential for achieving the resource shifts necessary to support the Department of the Navy's recapitalization goals. Chapter 4 provides the mechanism for introducing the strategy and enablers to the Department of the Navy. Supplemental information is provided in the appendixes.

2

Selected Navy Initiatives

The Office of the Chief of Naval Operations for Logistics (OPNAV-N4) is responsible for staff oversight for all facilities and installations of the Navy's shore establishment regardless of mission, and is the functional or mission sponsor for logistics (supply, maintenance, transportation) and base support (family housing, fire, and safety). Because this office sponsors a number of programs specifically with the aim of infrastructure cost reduction, the committee began its deliberations using this office's agenda as a roadmap. This chapter summarizes most of these initiatives and comments on progress in applying innovation in organization, technology, and management to the infrastructure as viewed in the first half of 1998.

REGIONALIZATION

Most Navy bases and facilities are clustered in a few regions around the country. Before regionalization, each facility was managed as a subordinate unit of a parent command and was supported independently of the other activities in the region, often providing all of its own infrastructure support. Over the years, the Navy has attempted to coordinate support activities in a region, but in general, these efforts have been limited. The Navy is now consolidating support in its larger U.S. regions, and it has termed that program "facility claimant consolidation" and "regionalization."[1] This section reviews regionalization and considers ways to improve the success of that initiative.

[1] In part this is due to base realignment and closure (BRAC).

Current Status

To understand the regionalization initiative and its implementation challenges, the committee visited San Diego naval bases reporting to the Commander in Chief, Pacific Fleet (CINCPACFLT). The San Diego region has the second largest concentration of Navy personnel and units, and it is home port to 56 ships. Given base closings elsewhere, it appears that the Navy presence in the region will increase over the next few years. Although most of the bases participating in this regionalization effort are in the immediate San Diego area, some bases, such as Fallon, Nevada, and El Centro, California, are further away. To capture this wider consolidation, the new region is called the Southwest Region, and it covers three states.

As part of the regionalization process, ten San Diego bases are first being restructured into three megabases. Most of the money to operate and maintain the bases will come through one claimant (CINCPACFLT). Several other claimants have transferred money and control of property to CINCPACFLT as part of the process to simplify the flow of funds, and most of the installation support money comes to CINCPACFLT under the sponsorship of OPNAV-N4. The committee did not obtain a full understanding of all the planned savings because regionalization was still in the early stages of development and implementation.

Regionalization will consolidate the support of the three new megabases. The regional command will have consolidated departments headed by program managers (or assistant chiefs of staff [ACOS]). The departments will provide a range of services to the three megabases, such as facility management, security, port operations, and air operations. There is a regional advisory board consisting of the megabase commanding officers and department managers.

Based on the study of seven functional areas, the initial consolidation should save at least $20 million when fully implemented. The goal is to achieve savings of $40 million a year. Most of the savings from regionalization will come from reducing the number of middle management civil service employees and overhead features. Prior to the consolidation, bases in the region had reduced costs by $30 million. This was a 10 percent reduction in their base operating support budgets. The cuts were the result of overall budget reductions—some were met through efficiencies and others through reductions in service. From 1997-1998, 500 civilian workers were displaced by these cuts.

It was very difficult to account for the total cost of the infrastructure resident in the San Diego region. Excluding personnel assigned to ships and to deployable aircraft squadrons, there are 36,000 military personnel and 20,000 Navy civilians in the region. The personnel cost alone is $2.5 billion. This estimate excludes construction costs, contractors, and utilities. As far as the committee could determine, the regional commander will have control over only $500 million. Thus the identified savings ($20 million) and the savings goal ($40 million) only begin to touch on total Navy expenditures in the region. Additional savings may be forthcoming, but as yet have not been identified or targeted.

The committee was also briefed on the Hampton Roads, Virginia, regionalization effort, which reports to the Commander in Chief, Atlantic Fleet (CINCLANTFLT), and has an organization and management matrix similar to those in the San Diego region. Consolidation of the Hampton Roads region, the largest naval concentration, involves 12 installations with 17 claimants, 25 property managers, and more than 500 tenants, with an inventory of $10 billion in buildings and 56 square miles of land. Although most of the examples below are drawn from the Southwest Region, the strengths and concerns with consolidation also apply to the Hampton Roads region. Figure 2.1 and Table 2.1 summarize the current status and projected annual savings for the Navy concentration areas in which regionalization and consolidation are to take place.

Rationale

Regionalization is a form of consolidation. In theory, consolidation can either improve or degrade the use of resources. However, there are many reasons to expect lower costs, i.e., scale economies, scope economies, redundancy elimination, and market leverage.

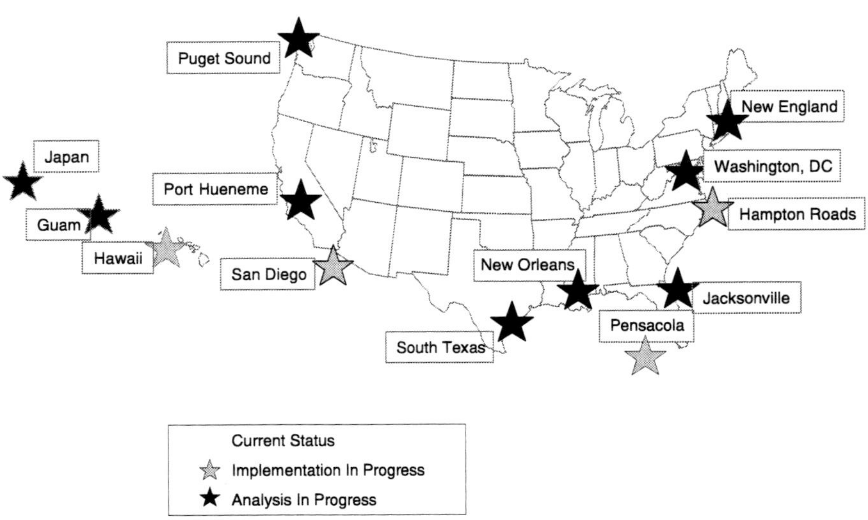

FIGURE 2.1 Navy concentration areas.
SOURCE: Adapted from Wennergren, David M. 1998. "Navy Concentration Areas—Host Activities," *Navy Infrastructure Cost Reduction Initiative,* Office of the Chief of Naval Operations (N464), Washington, D.C., briefing to the committee, February 4.

TABLE 2.1 Current Status of Regionalization

Navy Concentration Area	Projected Annual Savings	Current Status
San Diego	$40 million	Analysis complete; PACFLT implementation in progress. Consolidating 10 hosts with regional base operating support (BOS) service delivery
Hampton Roads	$83 million	Analysis complete; LANTFLT implementation in progress. Consolidating 11 hosts with regional BOS service delivery
Pearl Harbor	$18 million	Analysis complete; PACFLT implementation in progress. Consolidating 8 hosts with regional BOS service delivery
Pensacola	$15 million	Analysis complete; CNET implementation in progress. Consolidating 4 hosts with regional BOS service delivery
Pacific Northwest	TBD	Analysis in progress
Jacksonville	TBD	Analysis in progress
Washington, D.C.	TBD	Analysis in progress with N464 support
Guam	TBD	PACFLT regionalizing as part of regional A-76 competition. Consolidating COMNAVMARIANAS and NAVACTS Guam, regionalizing BOS functions
Japan	TBD	Analysis in progress with N464 support
Port Hueneme	TBD	Analysis in progress; some functions being implemented
New Orleans	TBD	Analysis in progress with N464 support
South Texas	TBD	Analysis in progress with N464 support
New England	TBD	Analysis in progress with N464 support

SOURCE: Adapted from Wennergren, David M. 1998. "Regionalization: Current Status," *Navy Infrastructure Cost Reduction Initiative,* Office of the Chief of Naval Operations (N464), Washington, D.C., briefing to the committee, February 4. TBD, to be determined.

Scale economies come from consolidating common workloads. Basically, when the workload doubles, something less than double the amount of resources is needed. There could also be scale diseconomies, which would make it more costly to provide services in larger organizations. This could be caused by additional oversight to coordinate the work. For the type of work consolidated in the Southwest Region and the size of the workload, it is unlikely that there are scale diseconomies.

Scope economies are efficiencies gained from merging different types of workload. For example, there are scope economies from naval aviation depots (NADEPs) working on both airframes and components. These workloads may use common management or inventories. The two workloads may also complement each other. For example, warehousing and configuration management may be collectively less costly if they have a common provider of the services.

When each facility has its own supporting unit, there could be redundancy in the support. As an example, a public affairs officer at each base may be responding to the same question from the same community representatives. Because smaller bases have often reported to different higher-level organizations (claimants), there was limited consideration of joint provision of these redundant services.

Many apparent redundancies are probably scale economies that show an excess only after consolidation. For example, finance and accounting personnel collectively may be able to manage accounts with fewer people than the sum of the individual departments. There is no redundancy because all of the people are necessary to meet the needs of the individual bases. The point is that classifying the source of savings is somewhat arbitrary—although the gains are very real.

Sometimes, consolidation can improve the purchasing power of the new unit. This is often termed market leverage. Private firms may offer discounts to larger customers, in this case the regional service providers, and this would reduce the cost to the consolidated units. The saving derives from the outside service provider's own scale economy.

Regionalization need not mean that one central organization provides all the support in a geographic area. For example, it could be agreed that one base in an area would operate the bachelor quarters, another facility would handle security, and still a third facility could handle personnel and pay support services. In this case, the region would take advantage of scale economies in the individual services. However, there could be significant drawbacks to such a consolidation. First, it would be very difficult to trade off across support areas (i.e., it would be hard to move resources across support areas when there are unforeseen problems or unexpected changes in the demands for the services). Also, there would be no gain from scope economies. Any gains in consolidating management oversight across services would be lost. For the San Diego and Hampton Roads regions, the service providers would be part of a larger organization, and this should allow the shifting of funds.

Some Concerns and Suggestions

Understandably, some reluctance exists at local bases and facilities to participate fully in regionalization. In part, it is a feeling that commanders and department heads would lose control over the quality and level of service support.[2] There are many reasons for the Navy, regional commanders, and bases to be concerned with the outcome when consolidating service delivery. The newly consolidated organization could grow over time, particularly its indirect labor

[2]Also, the BRAC process may not be over, and if so, further adjustments will have to be made.

component, and the consolidated organization would have the challenge of properly prioritizing and sizing services at individual facilities. All of these potential problems can be addressed beforehand, but they cannot be ignored.

The committee identified several ways to improve the regionalization initiative, make it more responsive to the customers served, and guarantee that the savings are realized. These improvements are listed below and discussed in the following paragraphs:

- Simplify the organization.
- Charge fees for service.
- Establish full cost visibility.
- Integrate competitive sourcing.
- Fully integrate all bases in a region.
- Seek customer input, and measure customer satisfaction.
- Establish metrics jointly with customers.
- Institutionalize success.
- Better train and assign shore managers.

Some of the recommended changes would directly address local concerns of bases. Having customers pay for additional services would give them some control over the quality of services. Also, clearly defined metrics provide both customers and regional service providers with performance standards to identify performance problems and gaps.

Simplify the Organization

In the San Diego region the new organization appears to have been superimposed on the previous organization, which in turn appears to have been superimposed on previous organizations. The new program managers, or ACOSs, are base commanders or former base commanders. There are 20 proposed functional departments. Some may be merged into common departments, but a large number are still likely to report directly to the regional commander. A large regional advisory board was established to make major budget shifts across functional areas. Although all of this seemed to be necessary to achieve buy-in from the current organizations, this new organization may be hard to manage, making it difficult to make decisions and forcing customers to work with many departments to solve problems that cut across the 20 areas.

Experiences of other large complex organizations indicate that with the support of higher leadership, organizations can be streamlined and made more customer friendly. The number of departments should be relatively small, preferably between five and ten. A reasonable set of departments in the region would be ship support, aircraft support, personnel support, facility support, and safety.

Regionalization might be an opportunity to realign previously constructed regions. Although there is no reason to believe that the optimally sized regions are the same for all types of services, having different-sized and overlapping regions is not customer friendly and may hinder future consolidations. Previously regionalized activities that should be reexamined are maintenance, personnel support detachments (PSDs), NAVFAC's Southwest division, and housing.

In general, those in charge of facilities (or bases) should not run departments that are intended to serve many facilities. As currently envisioned in the Southwest Region, base commanding officers (including those at bases being merged into the larger units) will head the departments. This means that the commanding officer must allocate support across his/her base and other bases. As objective as they try to be, those in charge of facilities or bases have conflicting incentives and goals—keeping up service at their own base and properly allocating resources to other bases. This is an unnecessary tension in the system that can never be satisfactorily resolved. In some cases, such as port operations, for which only one or two bases currently perform the functions, the issue is less pressing. However, separating the jobs should be an objective.

Charge Fees for Service

Two methods exist for funding infrastructure-related goods and services: (1) fees for services paid by the customer, and (2) funding directly by the service provider. To control demand for support services, units could be charged for some additional services beyond those provided by existing Navy Working Capital Fund (NWCF) activities. This would enable the internal "customers" to signal which work and services are worth the costs. As currently planned in the San Diego region, only a few units will pay for facility services. This issue becomes more of a concern when a centralized unit, and not the individual facility, makes most of the decisions. As currently proposed in the San Diego region, program managers will have to petition the regional advisory board for funds. The program manager then basically becomes an advocate for his area. Each program manager can justify additional funding because customers will make requests without budgets and without paying for the services.

A drawback with establishing added fee-for-service activities is that they would most probably fall under the rules of the current NWCF. The current NWCF provides a much-needed flexibility, but it has lost the confidence of many customers. In addition, charging fees for services requires added administrative and accounting processing. Thus, decisions to charge fees for services should be made only when benefits will clearly outweigh overhead costs.

Establish Full Cost Visibility

Payment for service is related to cost visibility, but it is not the same. A good accounting system could capture the cost even without reimbursements.

The costs should be visible to both service providers and internal customers. In general, these costs have not been very visible to anyone. Part of the reason is that facilities do not have all resources in their budgets and indirect costs are not captured by anybody. The costs of military personnel and construction are often left out of the cost of delivering a service. Indirect costs of a base's service probably should include the support for military personnel, such as housing, personnel offices, and the other quality-of-life features provided on a base.

The regionalization plan in the San Diego region does include a program to introduce activity-based costing (ABC). When this program is put in place and maintained, it should be extended to capture all costs, including military personnel and construction. Currently, the San Diego region plans for ABC do not include the full cost of military personnel. Full cost visibility could help establish fees for service and help facilities manage their operations more efficiently.

Integrate Competitive Sourcing

The Navy, like all of the Department of Defense (DOD), has an aggressive initiative to open to competitive bidding many of its support services. The effort builds on the previous success of competitions over a wide range of activities and organizations. To date, the Navy has been performing these competitions at individual bases without the benefits of regionalization.

Regionalization and competition are complementary because competitions for larger activities generally produce a greater percentage of savings than do smaller competitions. Moreover, regionalization should reduce the administrative cost of competition, and the new regionalized departments should be able to represent the in-house bid in a more effective manner.

As suggested in Chapter 3 in the section titled "Competition," regionalized activities should be competitively sourced. The Navy should announce competitions as soon as it decides to regionalize, or as soon as possible thereafter. This would give private firms, as well as the in-house work force, time to restructure and reengineer activities. Industry forums could improve both in-house and outside bids. If, in fact, an in-house team is successful in a competition (as has been the case in the past [General Accounting Office[3] studies]), it provides direct market evidence that its organization is the most efficient.

---

[3]U.S. General Accounting Office. 1998. *Public-Private Competitions; DOD's Determination to Combine Depot Workloads Is Not Adequately Supported,* GAO/NSIAD-98-76, Washington, D.C., January; U.S. General Accounting Office. 1997. *Privatization and Competition; Comments on H.R. 716, The Freedom from Government Competition Act,* GAO/T-GGD-97-185, Washington, D.C., September 29; U.S. General Accounting Office. 1997. *Defense Outsourcing; Challenges Facing DOD as It Attempts to Save Billions in Infrastructure Costs,* GAO/T-NSIAD-97-110, Washington, D.C., March 12; U.S. General Accounting Office. 1996. *Defense Depot Maintenance; Privatization and the Debate Over the Public-Private Mix,* GAO/T-NSIAD-96-146, Washington, D.C., April 16.

Fully Integrate All Bases in a Region

As noted above, many of the naval bases in the San Diego region are reluctant to participate in the regionalization. In the San Diego region, the naval aviation depot (NADEP), Space and Naval Warfare Systems Command (SPAWAR) activities, and the hospital should participate fully. Some argue that regionalization should not include all activities in direct support of missions. Any attempt to separate activities on the basis of mission support reduces the merits of the consolidation. Mission-funded organizations should be directed to move as many of the support people into the consolidated organization as possible. By retaining budgets and purchasing services through a reimbursable system, mission-funded organizations would retain control over the level of services.

Seek Customer Input and Measure Customer Satisfaction

The current suppliers of services are basically shaping regionalization. It makes sense for functional experts in facilities, supply, air operations, and so on, to work on the design of the new organization. However, it is natural for the current providers to construct an organization that mirrors the individual, smaller organizations except for the elimination of some middle management positions. Also, there will be an inclination to continue to provide the services as they have in the past. Yet, the old organization may not have been satisfying its customers, and customers' inputs are needed to help shape and reengineer the new organization. As proposed in Chapter 3, the committee recommends that each of the new regionalized departments establish a performance work statement with output metrics based on both customer and provider inputs. These mission statements would help define the departments and the subordinates.

All of these concerns will exist in a decentralized system, where each base and facility makes its individual decisions. However, when individual commands are closer to the end users (or may be the end users themselves), they would be more aware of changing needs and local priorities. Often, centralized decision makers can consider priorities among functions, but they may find it difficult to make across-base comparisons. Consideration of customer inputs and fee for service are key ingredients in any part of the reorganization and reengineering process needed to offset the bureaucratic tendency of centralized organizations.

Establish Metrics Jointly with Customers

A common challenge for organizations is to define meaningful metrics for judging performance and to use these measures properly. The more centralized the decision process, the more important is the role of metrics. The consolidated service providers, often headquartered away from the base, have to judge the quality and adequacy of services. There are many layers and levels of metrics. The higher-level metrics reflect the overall functions and goals of the infrastructure, and the lower-level metrics are more clearly related to day-to-day resource decisions and performance measurement. The higher-level metrics provide context and direction to the organization. Regional commanders are generally not accountable for their magnitude and changes. It is the lower-level metrics on which they can be judged. The greatest challenge is to connect the lower-level metrics to the higher-level ones. Even when those connections are not well defined, and they often are not, it is still important to track both sets systematically. A "best practice" of many successful firms is to have a regular audit process that reviews metrics and publishes the results. Some examples of higher-level metrics are the following:

- Readiness condition of home-ported ships as they leave for deployments,
- Readiness condition of home-based squadrons as they leave for deployments,
- Retention rates of Service members in the region,
- Percent of regional Service members requesting to stay in the Southwest Region, and
- Customer satisfaction on surveys.

Each individual metric has its weaknesses. Some are better applied to a base than to a region. But collectively, they present a picture of what material and personnel support should do. Since regional support is only one component of success, some trends may be unrelated to regional performance. Yet, a downward trend because of outside forces may justify increasing or shifting resources by claimant and resource sponsors. Common metrics from the CNO to base departments should focus the organization on its mission and purpose.

Lower-level metrics are important for day-to-day regional management. The set of metrics selected should be constantly reviewed. The goal is to capture a reasonable set of indicators that collectively provide a picture to the regional commander.

Examples of some lower-level metrics are as follows:

- Number of hours an airfield is closed,
- Number of nested berthings at pier,
- Square-foot cost to operate and maintain buildings,
- Elapsed time for repairing problems in housing,
- Utility costs per square foot of space,
- Crimes committed at Navy installations,
- Elapsed time for remedying environmental incidents, and
- Waiting time for family services.

After the metrics are identified, the Navy must establish benchmarks. Benchmarks could be based on historical "good periods," the performance at the best installations, or where comparable measures can be made with private-sector performance.

Institutionalize Success

The Southwest regionalization effort is being built around the personalities of the region's leaders. The committee noted that they have built a partnership and a feeling of trust that allows the bases and tenants to break away from the traditional organizations. However, these relationships are fragile, and the rotation of key personnel could easily erode the gains. Over time, the regionalization may retain its form on paper, but a different set of leaders could revert to old practices and recreate their own support. The Navy must adopt policies and mechanisms of procedure and process to guarantee that successes endure and that process improvement continues.

Better Train and Assign Shore Managers

When organizations are stable, good general managers can come in and effectively run the operations. When rapid change is called for, leaders with a deeper understanding of how support organizations work and with enough tenure to make changes are required. The Navy must decide how to better train and assign its shore establishment leaders. These leaders must understand the component departments, needs of customers, current technologies, and modern business practices. The Southwest Region has a good share of experienced leaders, but this is unusual given current assignment policies and rotation. The Navy should consider extended tours for base commanders, require one previous base command assignment before assignment to the top position, and ensure promotion opportunities for excellent performance in shore command assignments. The executive officers could be Navy civilians with 5-year contracts. Such contracts allow for continuity and protect the Navy from poor performance. The Navy should cooperate with bases of other military departments in the same area. Joint efforts and cooperation in emergencies, safety, counterterrorism, and so on would be desirable, and likely necessary. Joint efforts in housing and environmental controls might further reduce costs for all concerned.

The Future of Regionalization

The Navy has made significant progress in overcoming parochial, "stove-piped" interests in favor of implementing regional cost-reduction strategies across multiple activities and major claimants. But more can be done. The Navy could expand its ongoing regionalization efforts to include other "ashore" funtional areas. Opportunities still exist for significant consolidation savings in the areas of training, information technology, and ordnance, as well as expanding and building on existing regional maintenance successes. The Navy could also continue to look for opportunities to reduce command structures, by minimizing the number of installation management commands within a Navy concentration area, building on such efforts as the merger of the Naval Amphibious Base, Coronado, with the Naval Air Station, North Island in San Diego, California.

The Navy must also continue to overcome the initial reluctance of some activities to divest themselves of personal control over a functional area, in favor of adequate access to regional service delivery organizations. The Navy will also have to continue to be vigilant to ensure that regional organizations continue to reinvent and reengineer themselves over time to avoid organizational inertia or growth.

Recommendations

- *Local commanders should follow the guidelines enumerated above in implementing regionalization directives within their regional bases.*
- *The Navy should take advantage of its claimant consolidation and regionalization initiatives to further reengineer its essential support activities.*
- *The Navy should initiate regionalization and consolidation programs throughout all second-echelon commands not affected.*

FACILITIES PLANNING, MAINTENANCE, AND DEMOLITION

Facilities management is identified as a key business function under the Navy's proposed business model for regional base commanders. It encompasses a wide range of activities to include (1) public works management covering installation planning, engineering, and real estate and acquisition; (2) all forms of utilities; (3) all forms of shore transportation; (4) facility maintenance and construction; (5) environmental functions including prevention, compliance, and restoration; (6) building services; and (7) grounds maintenance.

The physical shore structure consists of about 150 base installations, plus 200 reserve facilities. Each installation consists of a varying amount of class I (land) and class II (structures) real estate. In some respects, the sheer magnitude of the real estate investment would suggest a major opportunity for cost savings through closure, regionalization, consolidation, demolition, innovative usage, leverage through private-public ventures, and improved processes and management.

These challenges are being addressed vigorously by the Shore Installation Management Division (OPNAV-N46), the Facilities and Engineering Division (OPNAV-N44), the regionalization efforts in the San Diego and Hampton Roads regions, and by other areas picking up the challenge. Simultaneous approaches at various levels along various fronts may be a better way to maximize the benefits at a faster pace, but they must be integrated within the wider parameters of overall facility planning based on an optimum shore configuration to satisfy mission needs. A top-down commitment of support, a willingness to assume judicious risk, an integration with the strategic vision, and a dedication to implementation are necessary factors for success.

Real Property Maintenance

Despite a reduction in the number of facilities and the average age as a result of base realignment and closure (BRAC) plus an annual maintenance investment approximating $1.0 billion, there remains a critical backlog of maintenance(about $2.4 billion at the end of FY 1997. This backlog is projected to grow to $3.4 billion at the end of FY 2003 despite an expected increase in annual appropriations to $1.3 billion in FY 2003. The largest investment category with a critical backlog appears to be military housing. In the meantime, added deterioration will continue to increase the backlog. Disparities of such magnitude tend to force a focus on short-term solutions rather than on a long-term life-cycle approach, because overtaking the shortfall appears to be unreachable.

This critical backlog of maintenance and repair engendered a great deal of concern on the part of the committee, primarily the degree of confidence in its reality. The methodology used for evaluation and the data maintenance system are robust; however, they may produce inflated figures if there is no appropriate benchmark related to mission requirements for evaluating conditions. On the surface, having the ratio of real property maintenance (RPM) funding to remaining critical backlog hover at approximately 1:2.6 between FY 1997 and FY 2003 would lead to the conclusion that continued neglect could result in a collapse of the inventory. It is recognized that the projections presented reflect advancing deterioration, new finds (undiscovered at previous survey), and inflation, which complicates an analysis. It is also recognized that an accurate projection is tenuous and that extensive effort goes into making the projections. Field surveys for maintenance needs are conducted using centrally developed guides and standards developed to assist claimants in assessing the condition of facilities. These findings are then reviewed by higher authority to identify what is considered the critical backlog, i.e., that level at which lack of maintenance would impinge on operational readiness. Considering that the RPM funding allowances are far below the projected critical backlog, it would appear there is a lack of credibility as to (1) the actual impact on base readiness or (2) the validity of the condition assessments.

This problem suggests the need for a benchmarking system that would gain and hold the confidence of the budgetary and policy decision makers and eventually the U.S. Congress. To do this will require a fresh look at the total problem, with a focus on establishing reasonable standards of maintenance consistent with mission requirements, best business practices, and an automated information system that can be used for prioritization. This look should include a review of the approaches that large, private-sector facility owners and facility management providers take with respect to RPM and the strategies they employ to reduce costs, including automation of data. It would appear that benchmarks have to be established for the conditions of maintenance for various types of facilities, as well as for the effectiveness and efficiencies of utilization of various facilities. Ingenious measures are needed to effect savings, including alternative forms of acquiring the necessary maintenance; management strategies for reducing operating costs; and introduction of information systems for tracking conditions, costs, and meaningful metrical outputs.

Real Estate Ownership

At one time, NAVFAC was recognized as the owner of the real estate in the Navy. Now ownership has been transferred to fleet regional or base commanders. This approach has merit because it places responsibility for determining and satisfying requirements within the operator's control rather than within a service organization. In the case of regional commands, consolidating or reducing existing real estate requirements can be an effective objective for generating cost savings.

However, it has to be questioned whether decentralized ownership and distributive decision making are the ultimate answer to maximizing the potential for reducing the vast and varied holdings to the minimum needed to meet strategic requirements. On the surface, it appears that decentralization will lead to varying standards of occupancy and function, will not be integrated with overall Navy needs, and will more than likely result in the retention of more real estate than necessary. Some large industrial owners have traveled a similar path toward decentralizing the real estate function to respective business units but are now tending to look more favorably on "corporate ownership and management" of real estate. This has some advantages: it frees the business product manager from worrying about real estate (a non-core function), and it ensures that real estate holdings are more consistent with the corporate vision for the future and the standards of investment.

Sheltering the Force

Sheltering the force involves more than furnishing and maintaining the physical facilities for housing. It extends into the quality of life, appropriate housing allowances, off-base versus on-base accommodations, and government versus private-sector provision of services. Much of this function is policy driven and is not specific to a given installation, which indicates that major decisions will have to be made at the headquarters level for any changes to take root in the field. It is a program of major proportions: 31,500 units are judged by the Navy to be in need of major repairs; 11,200 families are considered to be unsuitably housed on allowances; and it costs the Navy considerably more than the private sector to own and operate housing. With a goal of improving housing conditions and providing higher housing allowances, the financial situation will be exacerbated. Therefore, additional savings will have to be extracted through other means such as smart public-private ventures, more effective maintenance and management, and incorporation of life-cycle costing. The successful leverage experienced in securing new housing units at Corpus Christi and Kingsville, Texas, through public-private ventures serves as a good example of innovative buying. Such ventures will become more complicated when existing units are incorporated. For the long term, the Navy should look to the private sector to house its personnel. Other Services, in trying to transfer their housing to the private sector, have had very mixed results. However, this is a separate issue from increasing allowances so that personnel can afford better private housing.

Centralized Demolition Program

Demolition is a means of reducing the physical infrastructure, which in turn reduces the attendant maintenance and repair requirements. It also decreases the real property maintenance backlog and frees land for other purposes. In an Office of the Secretary of Defense survey, the Navy notes that there are roughly 1,600 buildings involving 10 million square feet of vacant space eligible for demolition.[4] Structures other than buildings are also available for demolition. Although supportive records do not exist, the Navy estimates the annual cost of maintenance in a caretaker status at approximately 1.5 percent of the current plant value. This would amount to more than $20 million annually for the surplus structures surveyed. The Navy demolition program was expected to produce maintenance savings in eight years; experience to date indicates faster repayment. In addition to the monetary savings, ridding installations of unused class II property will improve the quality of life, enable the apportioned maintenance personnel (contractor and military) to be assigned to more productive tasks, and provide unanticipated intangible benefits. The committee believes that this program should be pursued vigorously.

Energy and Utilities

NAVFAC is reducing the cost of energy and utilities services through such means as third-party financing of energy-efficient equipment, energy awareness and training programs, privatizing utilities, and better assessing energy usage.

---

[4]Hollinger, CDR Thomas, USN. 1998. "Navy Centralized Demolition Program," Office of the Chief of Naval Operations (N443), Washington, D.C., briefing to the committee, February 5.

Utilities can be privatized, with sizable monetary and manpower savings. Promulgating satisfactory privatization agreements may require transferring real estate ownership. This and other security aspects will have to be addressed to create an efficient business climate. The committee views this area as a major opportunity for generating savings in facilities management.

Southwest Region Initiatives

Although the emphasis to date has been on regionalization and consolidation, other steps have been taken as well to generate infrastructure savings. The Civil Engineer Corps (CEC) officers in the San Diego region have been deeply involved and recognize their role as a staff support function. They also recognize the challenges yet to be faced in extracting further savings. One factor is the apparent lack of a benchmark to measure success. Establishing the traditional quantitative savings benchmarks in the facilities arena is not difficult; however, this alone does not necessarily measure quality or customer satisfaction. Metrics will have to be developed that are meaningful to users and will thus motivate them.

Southwest Division NAVFAC Regionalization

Although NAVFAC is no longer the owner of the real estate and facilities in the San Diego region, it is responsible for the staff support attendant to management of the facilities. Direct support is provided by the Public Works Center (PWC), with specialist support provided by the Southwest Division of NAVFAC whose area of responsibility encompasses a large portion of the western United States. The role of the Southwest Region in the San Diego regionalization was evident in three distinct areas: (1) shore facilities planning, (2) acquisition of construction and facilities services, and (3) the environmental function.

Centralizing the planning function for the region allows the establishment of a broader and longer vision for the area, provides the opportunity for the most efficient use of real property assets, enables establishing models using a larger database, and provides for more efficient and effective use of the personnel involved in the process. Planning over an extended regional base is necessary to ensure the optimum integration of facilities. Short of centralized planning, regional planning should be promulgated throughout the shore establishment. In addition to the direct savings in manpower and preservation of talent from a consolidated staff, greater savings should come from developing better functional units.

Regionalizing assets involved in contracting for facilities results in a more uniform contracting policy, better integration of contracting officers and support, and more effective acquisition. With the advent of more outsourcing and privatization, it is incumbent that those responsible for acquisition enhance their talents toward becoming smarter buyers. This can be better done with an interactive "critical mass" of individuals and an organization that has a stable workload. Performance has to be measured by the commands that receive the services and support to ensure that responsiveness and quality of service do not deteriorate.

In the case of the environmental function in the San Diego region, the ACOS for environment is the Commanding Officer, Submarine Base, with the Commanding Officer, Southwest Division NAVFAC serving as a deputy ACOS responsible for the cleanup program and the environmental aspects of construction, in addition to serving as a technical director for compliance. Another deputy ACOS is responsible for the normal air, water, and solid waste programs, with the PWC serving as the hazardous waste manager. This organizational structure provides centralized program management and a single voice with the regulators, and it is certainly more efficient and effective than if the problems were handled by the individual bases or installations. It may be advantageous to consolidate further as experience is gained with regionalization.

Real Property Maintenance for the San Diego Region

Prior to the regionalization effort, real property maintenance activities were largely accomplished independently by eight base entities. Today, many of those activities are concentrated at PWC and the Southwest Division NAVFAC. Regional-level functions are facility planning, space allocation, facility condition assessment, long-range maintenance plans, utility outages, regional prioritization, transportation management, and large contracts. To ensure responsiveness and tailored solutions, the base level retains responsibility for tenant liaison, maintenance management, emergency response, utility maintenance, transportation operation, and field contracts. Consolidation of the functions is projected to yield an upper and middle management savings of roughly 50 percent, or $7.5 million annually. This approach should also yield other economies of scale, standardize the level of service, and provide a central force to address potential savings areas over a broader range.

It should not go without mention that under the regional commander in the Hampton Roads region, a regional engineer has been established with the functional responsibility of life-cycle management (including determination of requirements) of the real property of the various installations within the command. This consolidation alone has resulted in a substantial savings in staff costs. Delivery of maintenance services by alternative forms (outsourcing or privatization) and by various competitive in-house suppliers is planned. The committee believes that the region is moving in the proper direction to achieve further savings and that its leaders are fully committed to the process.

Family Housing Privatization

Extensive efforts have been devoted by the San Diego region to looking to public-private ventures as a means of obtaining affordable housing for all members, eliminating the deficit in housing, and replacing or rehabilitating existing units. The studies of alternatives indicate a potential savings of close to 10 percent if some units are placed under a public-private venture.[5] Although a public-private venture does reduce the need for appropriated funds for capital construction, it also introduces a certain degree of risk against a marginal savings. This approach calls for a careful scoping of the contract and rigorous evaluation of the proposals to ensure quality and long-term benefits.

Regional Freight Transportation

A conceptual organizational plan and processes to manage freight movements within the San Diego region in support of the Naval Base Commander have been developed. They create a single Navy manager (Fleet Industrial Supply Center [FISC]) and two providers of service (PWC and Defense Distribution, San Diego, California [DDDC]) initially. This approach is projected to provide a monetary savings in freight costs of 36 percent and a manpower savings of 25 percent. These savings are to be achieved through relying on more scheduled deliveries versus on-call, and a more centralized and effective dispatch system enabled by a Transportation Information Management System (TIMS). Under the arrangement, DDDC continues to provide waterfront support and PWC provides vehicles and drivers to the regional transportation coordinator within FISC. The reason for having two providers instead of one, plus a coordinator, is unclear and warrants further study. There has been a revolution in logistics in the private sector. Many firms are outsourcing both their fleets and fleet management to third-party logistics providers. This activity may be a prime candidate for a sourcing competition.

Public Works Automation

Although the Naval Construction Battalion Center (NCBC) at Port Hueneme, California, has not yet been folded into the San Diego regional organization, there has been some self-imposed regionalization between NCBC and the nearby Naval Air Warfare Center (NAWC) at Point Mugu, California, that has produced impressive savings. Downsizing is accepted as a reality, as is the need to pursue further cost reductions. As a result, several programs have been initiated covering assets management, investment planning, energy, and automation.

---

[5]Usher, CAPT Jill, USN, Assistant Chief of Staff, Housing, Naval Base, San Diego. 1998. "Family Housing Privatization Concept for San Diego Region," briefing to the committee, April 27.

NCBC is the pilot Navy site for privatizing utilities. Its real property maintenance work is heavily outsourced; the option of bundling activities needs to be explored. The initiatives being carried out have the potential for significantly improving operations and for yielding cost savings; they should be further propagated and shared on a Navy-wide basis.

The showcase of NCBC efforts is the automation of the management of public works activities. This system, known as MAXIMO,[6] is built using commercial software products by blending them into a network system for facilities management with customer access. The MAXIMO system has reduced the automatic data processing staff, reduced lost work orders, reduced redundant work orders, reduced report generation, and increased "wrench" time for maintenance workers. The first-year return on investment is claimed to be 65 percent. MAXIMO is a step toward a more efficient paperless office and toward developing performance information that can be used to improve performance.

NCBC is also the pilot site for activity-based costing and activity-based management (ABC/ABM) implementation in the facilities management arena. ABC represents a cost assignment view and ABM a process view. It can be linked to the MAXIMO system. The benefits of the system are updated, reliable, outcome-based cost information for business decisions; timely and actionable information for management; and information for external reporting. Before proceeding with ABM, the ABC information should be used to determine if a sourcing competition is warranted.

Recommendations

- *Establish a central resource to provide guidance and standards, to promote innovation, and to ensure smart buying of services through the use of approaches such as public-private ventures and privatization of maintenance facilities.*
- *Establish a corporate real property management plan that integrates specific needs with strategic goals, force structure, and budget requirements; that determines occupancy standards and functionality requirements; and that operates on a fee-for-service basis.*

LOGISTICS

Early in its deliberations, the committee was briefed on several ongoing Navy logistics initiatives, many of which originated as early as 1993. These initiatives primarily centered on improving both the effectiveness and the effi-

[6]MAXIMO™ is the name of a software created by PSDI, Bedford, Massachusetts, to assist industry, government, and other organizations in maintaining facilities and production equipment.

ciency of the Naval Supply Systems Command (NAVSUP) and the various Navy activities involved in maintaining and overhauling its ships, aircraft, weapons, and electronic systems (which report to second-echelon organizations). Overall, the committee was impressed by the wide spectrum of activities proposed, which would require that the suppliers and customers of these services formulate mutually acceptable plans for improving performance over time. However, the results of such activities and their impact on reducing infrastructure costs are not yet visible, at least to the committee. A summary of the committee's review of the supply and maintenance initiatives is provided below.

Supply

NAVSUP is continuing to move toward a uniform, regionalized supply system partnership with the fleet by using the Fleet Industrial Supply Center (FISC) as the principal interface with the end user customer. Its vision is to develop a seamless system across several organizations whereby product lines are provided when and where needed. Samples of product lines include subsistence, equipment for individuals, tools, petroleum, lubricants, construction supplies, ordnance, repaired equipment, spare parts, and so on. The FISC would then become the focal point and provide all transportation and competitive sourcing activities in a geographical region, thus essentially regionalizing a "one-stop shop." The committee commends this initiative but notes that the methodological step-by-step, individual business case analysis used for reengineering each function across organizational boundaries and/or introducing the notion of competive sourcing without adequately considering bundling opportunities may not yield the desired result, especially when the scope of the individual emphasis is limited.

In addition, the committee is mindful of the challenges faced in modernizing and/or replacing legacy information systems to bring the Navy's inventory and supply management functions in line with the new vision. Even one of the most successful innovators from the private sector, Wal-Mart, has no moving customers, retail outlets, or warehouses comparable to the fleet, its battle groups, and supply ships. The committee believes that the Navy is vastly underestimating the scope and difficulty of a major integrated effort to reengineer and automate the critical functions of the supply management chain, especially when they involve the Navy's complex organizational and ship-shore interfaces.

Maintenance

A recent U.S. General Accounting Office (GAO) report on Navy regional maintenance states,

> The Navy has reported that more than $8.5 billion of Navy resources was applied in fiscal year 1996 to maintenance programs in support of fleet ships and aircraft. Each type of "platform"—surface ships, submarines, aircraft carriers,

> and aircraft—has a separate maintenance infrastructure. Maintenance is done at three different levels—organizational, intermediate, and depot—depending on the nature and complexity of the work required. Organizational maintenance is done by military personnel on board ships or at aircraft squadrons. While at sea, intermediate maintenance on large ships such as aircraft carriers and tenders is done by military personnel; ashore, intermediate maintenance is done by military and civilian personnel at submarine refit facilities and aircraft and shore intermediate maintenance activities. Depot-level maintenance is done mostly by civilian personnel at aviation depots and shipyards. In 1996, the Navy had over 21,000 military and 42,000 civilians participating in maintenance activities at the intermediate and depot levels. In addition, the Navy has reported that up to 40 percent of depot-level maintenance is outsourced to private companies.[7]

In March 1994, the Navy established its regional maintenance program to focus on reducing excess maintenance infrastructure. It was planned that the program be implemented in three overlapping phases during FY 1995 through FY 1999. According to the GAO and the Navy, however, the program is accomplishing other objectives such as improving maintenance processes, integrating supply support and maintenance functions, and providing compatible data systems across the three levels of maintenance.

With this background, the committee was briefed on the status of the regional maintenance concept by both OPNAV and the Southwest Region participants. The committee found that the organizational and reporting relationships among customers and providers of maintenance service have evolved over time and are indeed unusually complex.

In the aviation community, the fleet operators (in conjunction with air intermediate maintenance organizations normally at the air station) interface with the FISC for outsourcing all work not done internally, either to a prime contractor or to the naval aviation depots (NADEPs). The Naval Air Systems Command (NAVAIR) is the technical authority for all maintenance repair and overhaul activities of aircraft and controls a portion of the funds expended.

The provision of maintenance services for ships and submarines is handled quite differently: The port engineer acts as the general contractor attempting to integrate and coordinate the activities of several potential service providers including the shore intermediate maintenance activities (SIMAs), which are manned by military personnel, private shipyards, Navy shipyards, and the FISC. Although the Naval Sea Systems Command (NAVSEA) retains the role of technical authority for all classes of repairs, funding typically flows through the fleet.

Regionalization of fleet maintenance has the following noteworthy objectives:

[7]U.S. General Accounting Office. 1997. *Navy Regional Maintenance, Substantial Opportunities Exist to Build on Infrastructure Streamlining Progress,* Washington, D.C., November, pp. 1-2.

- One level of industrial capability;
- One owner ashore for industrial-level maintenance;
- Up to as many as eight regional repair centers, which are level-loaded with the overflow going to the private sector;
- A civilian and military work force integrated so as to take into account the protection of sea-shore rotation billets; and
- Common support functions across the regions from the view of funding, information management, management of transportation, and material—namely, a single point of contact for all maintenance activities.

The regional maintenance program has rationalized the role between the systems commands and fleet as follows: the fleet should deliver platforms to the CINCs. The fleet also should (1) control the industrial infrastructure and run these regional repair centers and (2) remain operationally in charge of commanders that run the maintenance operations. However, the systems' commands should remain as the technical authorities and be responsible for the life-cycle improvement of maintenance through acquisition and modernization management.

The committee believes that these goals, although laudable, need to be accompanied by additional actionable plans that are tied to achieving specific results over time. In addition, the committee agrees with the GAO findings[8] that the following challenges have to be addressed successfully before the objectives will be realized:

- In parts of the Navy, parochial and institutional resistance exists to the regional program objectives—the greatest resistance is to the elimination of organizations, the reduction of jobs and promotion opportunities, and the reduction of individual command's or organization's control over resources.
- Management visibility of maintenance-related costs, including the cost of military personnel, is lacking.
- Multiple, unconnected management information systems exist and do not provide adequate data for regional maintenance planning or decision making.
- Larger numbers of shore positions are desired to support sea-shore rotation than needed for maintenance personnel to actually perform the work.

The committee believes in the singular importance of the Navy having the ability to maintain fleet readiness and to repair damage expeditiously to ensure adequate combat capability. (The aim must be to provide the logistics capability at the right place at the right time.) Being effective in crisis and war is essential. The difficult balance to achieve is that of reducing peacetime infrastructure

[8]See note 7 and related text.

costs—that is, those incurred in periods of relative tranquility—while not creating unacceptable risks for crisis and war. The committee is of the opinion that further opportunity exists for achieving savings in the maintenance infrastructure through reengineering across existing maintenance levels (depot, intermediate, and unit) and recommends continued exploration of opportunities consistent with the level of risk that can be tolerated for crisis and war situations.

Recommendations

- *Provide regional maintenance coordinators with more authority than just "the power of persuasion" among equals to meet the regionalization of fleet maintenance objectives (and overcome some of the parochial Navy challenges listed above).*
- *Include the costs of the military personnel involved in maintenance and other logistics functions in information made available to decision makers at all levels. Without an understanding of true costs in the Navy, decisions will be made on the basis of intuition and incomplete facts.*
- *Continue to push for and eliminate excess capacity where it exists. The committee acknowledges external pressures inhibiting base closure and further contracting of depot work, but overcoming these pressures should not be abandoned.*

SMART BASE

Smart Base is a specific Navy project to test and evaluate initiatives that hold promise for reducing costs and/or improving the delivery of services on naval bases. Initiatives could be built around either recently developed technology or modern business practices; however, the focus of most of the project is on the use of computers to improve base support, with a common theme that computers more quickly and accurately complete personnel, financial, and facility transactions. Some of these initiatives rely on new software for individual computers, but most of the applications require that computers be linked and pass information. As the committee notes below, this lack of connectivity among functional computers has been a major limitation of these initiatives. Other initiatives, such as Smart Card, examine innovative ways for the Navy to use evolving commercial standards in embedded-chip credit cards, much like those the commercial banking industry has attempted to provide to its customers. Essentially, the Navy's Smart Card would enable electronic access to personnel records for record keeping, security clearances, travel reimbursement, and so on.

As part of its review, the committee visited the Naval Station at Pascagoula, Mississippi, and the Portsmouth Naval Shipyard in Portsmouth, New Hampshire, which are principal installations testing Smart Base initiatives. When the committee visited Pascagoula, it was home port to four ships, 1,600 military personnel, and 90 civilians. Since then, the base has added a cruiser and will soon have a Coast Guard cutter. It is a modern and relatively small operational base.

Portsmouth, on the other hand, is an old and large industrial facility. It is one of the four remaining public shipyards, and it is dedicated to the repair and overhaul of nuclear submarines. The facility is located on an island with four million square feet of facilities, some of which date back to the Civil War. More than 90 percent of its current labor force of 3,400 are government civilians. Although the primary activity of the base is ship maintenance, about one-sixth of the work force, more than 550 people, provide base support.

Results to Date with Information Technology

The Smart Base project is 2 years old. In FY 1997 and FY 1998, the Navy will spend up to $20 million on its Smart Base initiatives. The Navy uses Broad Agency Announcements, which invite private firms to submit Smart Base proposals. To date, 200 proposals have been received—the largest number dealing with the use of Smart Cards; currently, 21 different initiatives are being tested at Smart Bases. For example, Pascagoula is the testbed for systems that support distance learning and bachelor quarters' management. The major projects at Portsmouth include an environmental information management system, an energy management system, and an electronic security system.

Although the project has not yet achieved any breakthroughs, it is too early to judge the merits of individual initiatives. A well-publicized failure at Pascagoula was the establishment of an automated kiosk that would have allowed personnel to check onto the base, check into the bachelor quarters, and update personnel records. Service members would use Smart Cards to carry out the transactions. The kiosk never worked and was recently removed. However, a Smart Card seems to have been used successfully on board the Navy's Smart Ship. It is noted that a version of the Smart Card has been abandoned most recently by a number of major banks,[9] but a few are continuing to use the approach. This seems similar to the Navy experience overall.

Lessons Learned

Pilot initiatives established to demonstrate successful innovation in day-to-day operations are difficult to implement. The committee is concerned that many of the current initiatives will not succeed and meet Navy objectives. Some of the shortfalls are listed below.

---

[9]In a recent newspaper article, it was noted that "use of the cards by customers had not come up to expectations and that merchants had pressed for a simpler operating system" (Authors, John. 1998. "Lack of Interest Forces Banks to Abandon 'Smart Card' Plans," *The Washington Times,* November 4).

Lack of Connectivity

Lack of information connectivity and networking among functions is a problem throughout the Navy—and the Smart Base experiments suffer as a consequence. One of the initiatives is to develop Internet access. Other initiatives basically assume, rely heavily on, or have limited value without the connectivity of computers. As a simple example, Pascagoula's consolidated bachelor quarters has five computers in its front desk area, each doing part of the room management function. Information cannot be passed between the computers, and all of the information cannot be displayed on a common screen. Another example: a particular software project allowed the purchase of office supplies electronically, but its design did not recognize that the task is normally done by clerks without Internet access.

Failure to Truly Use the Private Sector

An unspoken principle underlying the Smart Base project is that inefficiencies result from the lack of applying current private-sector technology. The thought is that putting this technology on bases will bring efficiency and service quality closer to that used in the private sector. Unfortunately, this approach could fail because what the Navy lacks in many areas is more than technology; it lacks the private sector's ability to integrate systems in order to solve business problems. The private sector sharpens these skills continually in a competitive environment. By acquiring only the technology, the Navy is underutilizing the private sector. For every new hardware or software feature acquired, the Navy will have to solve new problems to make the new technology work.

Poor Planning for Phasing Out Legacy Systems

The committee was told that essential data are still stored and transactions still processed on old (legacy) systems. Interacting with different functions and old systems requires people familiar with the operations and maintenance of those systems. One example cited was the recent arrival of a UNIX machine at Pascagoula for the maintenance of personnel records. The base had no support and training for the system, and there was no connectivity provided with other systems on the base.

Lack of Fleet Support

OPNAV sponsors and Washington program managers appear to work directly with the base, bypassing both the fleet commander and the regional commander. This practice has made the claimant leadership somewhat indifferent to the program because they have no stake in its success.

Poor Metrics

Like other public organizations, the Navy lacks profitability as a metric to motivate the best introduction and use of technology. That does not mean that it cannot focus on reducing costs and improving performance. However, with the new technology, there has been little or no attempt to introduce cost visibility to track the costs, or to introduce and track performance metrics. Although there is an initial analysis of return on investment, the Navy does not appear to have a follow-up program.

Lack of a Transition Plan for Successful Innovations

Overall, the Smart Base project is evaluating many innovative ideas. However, the committee believes that the project will have only a small impact on infrastructure costs because there is no overall plan to extend successful Smart Base initiatives across the Navy shore installations.

Other Efficiency Projects at Pascagoula

Pascagoula appears leaner than other bases, but this result has little to do with the formal programs because the base was originally set up without many support activities. For example, the base does not have family housing, commissary, club, library, or chapel—these services are provided in the community. The base is very modern; it employs fiber cable throughout, permitting connectivity for most purposes. The base management has also contained, reduced, and avoided costs in a number of ways, and three projects are worth mentioning.

Electric Steam Plant

One of the most impressive technological introductions on the base is the steam plant on the pier. The modern plant was built by Mississippi Power at far less than the Navy's planned cost, employing sensors throughout, with the monitoring and regulating of the system highly automated. The system does not require an operator. This plant and the way it was acquired involved discussions with the private sector, which had the incentives to integrate several technologies into a modern and efficient plant. The base management estimates that this approach has saved the Navy $1.3 million in construction costs ($2.3 million spent instead of $3.6 million) and $500,000 annually in operation and maintenance costs.

Galley

Another action that is saving money is the elimination of the base galley. The bachelors now receive a basic allowance for subsistence (BAS) and have small refrigerators and microwaves in their rooms. According to the base commander, this arrangement appears to be well accepted. Also provided is a Navy Exchange and a small McDonalds restaurant, and negotiations are underway with the Subway fast-food chain. The galley had cost $700,000 a year to operate, whereas the BAS cost is $200,000 a year.

Grocery Store Discounts

The base at Pascagoula does not have a commissary. The base management asked local grocery chains to provide discounts to active and retired military families, and two chains offer these discounts. The committee notes that there are hidden savings here. DOD currently provides a 20 percent subsidy to the commissaries, which in turn provide only a 10 percent discount to sailors. The private grocery stores are providing a 6 to 7 percent discount at no cost to DOD. Although Navy management has been unsuccessful to date in having the state waive sales tax on grocery sales to military personnel, the initiative is noteworthy.

Recommendation

Both the Office of the Chief of Naval Operations (OPNAV) sponsors and individual participants associated with Smart Base initiatives need to have regional commanders assess the likely importance of these technology demonstrations, and have them support transition mechanisms for those deemed to have high payoff.

INFORMATION TECHNOLOGY FOR THE 21ST CENTURY AND SMART LINK

Currently, there are two rather separate information technology (IT) initiatives in the Navy: (1) IT-21 is primarily for the fleet, and (2) Smart Link is for shore installations. While conclusions are offered in this section, recommendations within this area are given in the next chapter in the context of the committee's vision of a Navy-wide information plan.

Background

Information and Technology for the 21st Century (IT-21) is both a new concept and a set of commercial standards providing the framework for the evolution of IT in the Navy. The Space and Naval Warfare Systems Command (SPAWAR) has been designated the lead organization to move the Navy's current IT resources from a set of disparate information assets to a fully integrated system providing connectivity across the entire Navy. IT-21 includes the underlying foundational elements of electronic communication—among them metropolitan area networks (MANs), local area networks (LANs), personal computers (PCs), and backbone and related assets—that will advance the Navy's objective to function as a single, highly efficient enterprise. In a larger DOD-wide framework, IT-21 will contribute to the information superiority envisioned in *Joint Vision 2010*.[10] Coordination of interfaces with other Armed Services and joint programs will be a continuing challenge.

Within OPNAV, the Director for Space, Information Warfare, Command and Control (N6) is the resource sponsor and is responsible for implementing the IT-21 Naval Virtual Intranet (NVI). Under the current plan, these funds will be applied first to support fleet claimants, numbering some 120,000 personnel. Selected portions of the shore establishment will be upgraded as part of this initial endeavor. If this approach proves successful, then other Navy shore elements will be brought into the upgraded IT structure.

A dialogue with industry is underway on IT-21 architecture. In particular, the concepts for the NVI have been disclosed to industry, and industry has communicated back to the Navy its view for implementation. The NVI process is planned to result in major acquisitions, with an emphasis on maximum use of commercial off-the-shelf (COTS) products, use of commercial standards, and the use of desired performance rather than detailed Navy design specifications. Similar efforts by other Services have a similar emphasis on COTS products, which can facilitate efficiencies in linkages that will have to be made.

The Navy has also been evolving, separately from the other Services, the so-called Smart Link. Smart Link is a demonstration program, and some parts of the Navy have a vision that it should encompass many of the intended features of IT-21. Smart Link is designed to provide long-haul communications, linking shore installations throughout the country, operating across a commercial asynchronous transfer mode (ATM) backbone on the public switched (unprotected) network. Smart Link could also grow at a later point to provide support services afloat. The relationships between Smart Link and IT-21 are unclear to the committee, and the issue of transitioning legacy systems, in general, will require high-level Navy attention.

Information Technology Responsibilities and Authorities

The recent Navy focus on electronic connectivity, and particularly connectivity for the fleet, places SPAWAR in charge. However, the responsibilities for infrastructure connectivity are less clear. Moreover, control of information content, integration, collaboration, and sharing of infrastructure information are important functions, but it is not clear to the committee that any one individual, or any designated set of individuals, is in charge of them.

---

[10]Chairman, Joint Chiefs of Staff. 1997. *Joint Vision 2010*. Washington, D.C., July 26. Available online at <http://www.dtic.mil/doctrine/jv2010/jvpub.htm>.

The shore establishment employs numerous IT systems, supporting diverse functions and commands; these systems are not compatible with one another and are costly to maintain. Smart new software systems are being introduced under the new base regionalization concept, but there is no clear process in place to ensure that all bases will benefit from these regional developments. High-data-rate service exists, to a large degree, at Navy bases throughout the country with emphasis now placed on the LANs, MANs with associated servers, and PCs, which still need to be funded and installed. If regional IT "islands" are substituted for functional local IT "stovepipes," the problem of redundancy and lack of interoperability will remain.

Information Technology Standards

A number of organizations and functional groups are active in addressing Navy IT issues; chief among them the Chief Information Officer (CIO) of the Navy. The CIO is a critical element of the Navy's hierarchy in establishing IT policy and providing Navy-wide leadership. The CIO chairs the Department of the Navy Board of Representatives, composed of senior IT representatives from throughout the Navy, and speaks with authority in furthering IT interests. The committee notes that in the highly important area of standards, the CIO recently issued a new IT standards document that is now being reviewed by industry. Industry comments will be considered before this document is issued formally. The committee believes that there is no more important area than standards if the Navy is to maintain the necessary discipline to avoid redundancies and to achieve interoperability throughout its emerging IT infrastructure.

Understanding Information Technology Costs

From an industry perspective, the GartnerGroup says that the total cost of ownership for a personal computer is $10,000.[11] During its review, the committee was unable to obtain reasonable estimates of Navy IT costs. The Navy's IT, and the overall DOD IT for the defense infrastructure, have evolved over decades, with no architect responsible for pulling them all together into a coherent integrated system. The Bandwidth Baseline Assessment Memorandum (BAM) study,[12] done for the Navy fleet, has no known counterpart to document the

---

[11]GartnerGroup Total Cost of Ownership Research Team. 1998. *TCO (Total Cost of Ownership) 1: A Business Yardstick*, conference presentation, GartnerGroup, Stamford, Conn.; Redman, B., W. Kirwin, and T. Berg. 1998. *Managing Distributed Computing*, R-06-1697, GartnerGroup, Stamford, Conn.

[12]Cebrowski, VADM Arthur K., USN, Director of Space-Information Warfare. 1998. Memorandum for RADM Jay Yakeley, USN, Director of Programming Division (N80), re: POM-00 C4I Bandwidth Baseline Assessment Memorandum (BAM), Office of the Chief of Naval Operations, Washington, D.C., March 18.

current installed IT systems on land. The Navy does have a five-phase plan to identify the approximately 40 percent of IT needs for shoreside activities, but it will not include items bought with Navy operations and maintenance (O&M) funds. The committee could not find a good baseline inventory of machines, software applications and software systems, and connections to LANs and wide area networks (WANs). Without a blueprint for existing Navy IT infrastructure and a comparable plan or statement of needs for the future, the committee could not make an intelligent assessment of the overall costs for the Navy IT infrastructure, or projected IT budget needs. Parenthetically, this lack of information on total available IT assets also has caused a problem in assessing the extent of the year 2000 (Y2K) problem.

Navy Culture, Values, and Resource Allocation

There is a perception that when resources are scarce, the Navy places its priorities on platforms and weapons systems. Traditionally, IT capabilities and IT skills have not been valued highly. Programming large sums of money for IT often results in the funds being a target for budget reallocation.

The payoff from IT is not well understood. Moreover, with no central management of IT, there are limited or no perceived incentives to share, use best business practices, learn from each other, keep track of what hardware and software configurations are used, track how much is spent on personal computers, or obtain the total cost of ownership. There is limited insight into the costs of doing work (e.g., military personnel are viewed as cost-free in many infrastructure areas), and the value of additional shared information for management is not fully appreciated.

The Navy now is showing its understanding of the need to share fleet information as it changes from platform-centric to network-centric warfare. However, this need to share information to improve effectiveness has not yet permeated the Navy infrastructure.

Connectivity and Content

In the IT world, connectivity is the highway, sufficient bandwidth makes it a superhighway, and the hardware and software are the enablers. Information content is the essence of the value in the product—getting the right information to the right people at the right time. Metcalfe's law asserts that the power of a network is proportional to the square of the number of nodes in the network. Connectivity and access to shared information increase the value of the product to the customer. To date, the Navy has focused on laying the highway and increasing the bandwidth to address the issues of the fleet—the fighting force of the Navy. As the committee studied the essential non-deploying shore support to the fleet, it found that in the Navy IT world, the shore support has not had high priority.

Lessons from Industry

In industry, once the highway is laid, getting the right product to market ahead of the competition becomes the issue. Time is an important metric or measure of success. To reduce time from the process, non-value-added work and transactions have to be eliminated. Thus, management needs to understand what work is being done, by whom, and where. In large organizations, most of the information on what and how work is done is hidden. The costs of getting work done are also not transparent.

The transformation of U.S. industry, which has reestablished its strength in the global marketplace, is due to restructuring the way work is done and how information is shared. IT has been a major enabler of this change.

Tedious and time-consuming tasks have been reduced, or made easier to perform. With the advent and broad use of the World Wide Web (WWW), order and delivery processes have been revolutionized. Paperless acquisition and electronic commerce are thriving. The facility maintenance for millions of square feet can be managed by a handful of people, centrally located, using an intelligent database and decision support tools. Adult training and education are being transformed by sophisticated, engaging, multimedia courses designed for learning. Universities are now managing their operations online. The Western Governors University is a collaboration among 16 states to provide learning opportunities online—shared and run across state boundaries. The University of Phoenix has no campus; all communications take place online, and non-Web courses are given in rented warehouses or storefronts. Business is being done differently.

Making IT pay off in industry has been possible because of process engineering, collaboration, and sharing of information.

Conclusions

As in the private sector, information technology will become integral to the Navy's infrastructure. A culture shift for the Navy involves moving from a position of information control to one emphasizing the rapid sharing of information to enhance mission accomplishment. Specifically:

- The Navy has pockets of excellence in information technology, and applications thereof, but the knowledge is not shared well. Many separate IT initiatives are not working in concert.
- Information systems used in business and in the Navy infrastructure do not in themselves reduce costs (i.e., they rarely result in an overall net reduction

of personnel simply by automating capabilities). Information systems can be a lever for reducing costs by enhancing access to and accountability for services.

- For the infrastructure and shore support areas it is not well understood by those in the Navy, or clear to the committee, who is responsible for providing desk-to-desk connectivity.
- The Navy has not fully identified the means for or funded the complete desk-to-desk connectivity, including access to sufficient bandwidth, needed to enable essential users to share data as they can now share voice communication. The result is that the benefits of full connectivity are not being realized even though a major portion of the costs are being incurred.
- Funding Y2K solutions "out of hide" will divert funds and attention from achieving the cost-effective benefits of full connectivity across the Navy infrastructure.
- In the area of infrastructure information content, there is not a clear delineation of responsibilities, authorities, and accountability in the Navy.
- Output measures of performance for infrastructure business areas are lacking, and they will be key to defining the content of the information system needed for integrated management of the Navy infrastructure.
- As the Navy changes its warfighting focus from platform-centric to network-centric warfare, it may also be appropriate to modify its management of the manning, training, and equipping of its forces (i.e., realign assignments of responsibility and accountability for portions of the infrastructure) to better match what fleet commanders need from the shore support establishment and to improve the cost-effectiveness of the infrastructure.
- A precondition to achieving the full benefits of IT across the Navy, and particularly across the Navy infrastructure, is visible and committed leadership from the top. The Chief Information Officer (CIO) and the Director for Space, Information Warfare, Command and Control (OPNAV-N6), for example, have only the power of persuasion to influence Navy IT evolution.

CLOSING COMMENT

The magnitude of the cost savings needed to recapitalize the Navy at levels of $3.5 billion to $5.0 billion annually cannot be obtained simply by "working harder and smarter" on the $13.5 billion portion of the infrastructure focused on by N4 (i.e., base operations, installations, central logistics, and quality-of-life activities) and discussed in this chapter. Moreover, Navy information technology initiatives are focused predominantly on fleet rather than infrastructure activities. Therefore, information technology currently is not being utilized to its full potential across the entire Navy.

The committee could not find within the Navy a bottom-up process for achieving the 10 to 20 percent or greater infrastructure cost savings required because of the factors enumerated herein. Although progress made at individual units and commands shows that considerable savings are possible when cultural resistance and parochialism give way to modern business reengineering processes, such action has not been seen across the entire infrastructure. The next chapters treat the implications in greater detail.

3

A Strategy for Managing the Infrastructure

THE STRATEGY

Although highly structured, formalized strategic planning activities, with accompanying volumes of data and staff analyses, are no longer in vogue in the private sector, successful enterprises now do implement explicit, well-coordinated plans for action involving all functional departments, with a common set of goals and coordinating mechanisms. The major finding of this study is that the Navy lacks an overall strategy for achieving the resource shifts necessary to support its recapitalization needs.[1] This chapter draws extensively from experiences in both the public and private sector and offers a strategy for managing the entire infrastructure.

---

[1] The committee's finding that the Navy lacks an overall strategy is similar to a conclusion reached by the GAO in its investigation of the Department of Defense infrastructure. See the following reports: U.S. General Accounting Office. 1997. *High-Risk Series: Defense Infrastructure,* Letter Report, GAO/HR-97-7, Washington, D.C., February 1, p. 10, available online at <http://www.access.gpo.gov/cgi-bin/getdoc.cg?dbname=gao&docid=f:hr97007.txt>; and U.S. General Accounting Office. 1997. *Defense Budget: Observations on Infrastructure Activities,* GAO/NSIAD-97-127BR, Washington, D.C., April 4, p. 32, available online at <http://www.access.gpo.gov/cgi-bin/getdoc.cgi?dbname=gao&docid=f:ns97127b.txt>. Furthermore, in the first of these reports the GAO noted the following: "However, breaking down cultural resistance to change, overcoming service parochialism, and setting forth a clear framework for a reduced infrastructure are key to avoiding waste and inefficiency. To do this the Secretary of Defense and the service Secretaries need to give greater structure to their efforts by developing an overall strategic plan."

What Is to Be Achieved?

The objective is to provide essential fleet support at reduced cost to make resources available to recapitalize, modernize, and increase the performance of the fleet, and thereby strengthen the Navy's core capability (fielding and using fighting forces) for the next century. The current budget for the infrastructure is $26.7 billion, from which cost savings of $3.5 billion to $5.0 billion are required.

How Is It to Be Achieved?

The objective can be achieved only by changing the Navy's business structure and modes of providing operational support—specifically, by adopting modern enterprise process technology and practices; competitive sourcing of as many of the support functions as possible with acceptable risk; learning by doing, so that adjustments can be made to these unfamiliar activities along the way; and building in safeguards against loss of support for the core capabilities during the transformation.

The following elements are essential to a successful strategy for achieving the shifts in resources needed for recapitalization:

- Knowing the cost structure of all the infrastructure activities, so that the Navy can understand where the leverage is, and therefore where to expend its efforts;
- Setting goals (money and time), assigning consistent responsibilities and authorities, and making plans that can be followed;
- Setting up a management information system to track support-system performance and costs;
- Rationalizing command and organizational responsibilities, authorities, and accountability to accord with the overall objective of the strategy;
- Preparing revised internal rules and procedures, and proposing legislative changes;
- Involving those affected inside the Navy, in other Services, and in civilian communities; and
- Obtaining agreements and legislation as needed in the Department of the Navy, DOD, the Office of Management and Budget (OMB), and Congress.

With What Is It to Be Achieved?

The resource and leadership requirements to accomplish the strategy are discussed explicitly in this chapter. These requirements involve the following:

- Commitment on the part of the top levels of Navy command and administration to reach the objective, as discussed in Chapter 4; and

- Adoption of key policy, technology, and business management enablers for systematic use throughout the Navy, as enumerated below.

Findings

Efforts to date to reduce the cost of the infrastructure (achieved through regionalization, consolidation of functions, personnel and facility downsizing, competitive sourcing, Smart Base initiatives, and so on) have achieved only a small fraction of the resource shifts required to meet the Navy's modernization goals. The committee estimates that these efforts, when fully implemented Navy-wide, could reduce infrastructure costs by about $0.5 billion per year. However, the funds required to modernize the Navy are in the range of $3.5 billion to $5.0 billion per year. These current efforts are, therefore, inadequate to meet the modernization goals. A major evaluation and process change in the way the Navy conducts *all* elements of its infrastructure operations will be necessary to achieve the added cost reductions required to modernize the fleet. However:

1. The Navy currently does not have a proven methodology for conducting this evaluation or for implementing and managing change in its infrastructure operations.
2. The Navy also does not have a methodology for determining, on a continuous basis, current industry "best practices" for fleet support functions.
3. Networked information systems can be a major enabler for cost reduction by enhancing access to and accountability for services. Compatible software applications and systems and shared information elements that facilitate interoperability for management control must be available, in addition to connectivity, before the benefits can be realized.

The committee also believes that the scope of this evaluation and process change, sometimes referred to as "reengineering the organizations," needs to be broad enough to materially affect all of the major claimants for the infrastructure dollars in question and should go deep into the hierarchy so that responsible managers will have the opportunity to change both the quality of service provided and the cost of delivery. All this must be accomplished within a decision-making framework where the interests of all users of infrastructure products and services are represented.

The committee notes that no single staff element in OPNAV or the Secretariat has sufficient oversight responsibility and authority to manage all of the infrastructure resources highlighted in Figure 1.2. Thus, responsibilities and authorities to implement change must be made clear and issued by the CNO.

Recommendation

The senior leadership of the Navy should immediately initiate a process to improve the business operations of the entire Navy infrastructure with the goal of reducing costs by the $3.5 billion to $5.0 billion per year required to fund modernization of the fleet. To effect needed changes, a proven management methodology should be used that includes the following essential elements:

- *Establishment of a practical vision of the future, with quantifiable goals, shared by all of the senior stakeholders in the Navy;*
- *Development of strategic proposals to move in the new direction and to deal with obstacles;*
- *Development of tactical plans for implementation, including task forces, milestones, and time tables;*
- *Clear assignment of responsibilities, authorities, and accountability;*
- *Follow-up progress reviews at regular intervals;*
- *Development of key output metrics for success, tracking of accomplishments against them, and requirements for results;*
- *Incentives and rewards for positive performance and punishment for non-cooperation;*
- *Celebration of victories; and*
- *Continuous improvement, because change is continuous and cannot be accomplished instantly.*

This same methodology should be employed at successively lower levels of command to align their goals, incentives, and metrics with the Navy's larger strategic goals until the entire process of change has been institutionalized across the Navy.

NAVY-WIDE ENABLERS

Although the list of business practices that can be used to improve productivity and reduce costs is extensive, the committee recommends that the Navy focus on three key enablers for application throughout the infrastructure and that it make the commitment to implement them fully without exception and without compromise. The three enablers, discussed below in the context of improving Navy infrastructure productivity, are as follows:

- Performance measures, cost management and allocation of resources,
- Information technology for infrastructure management, and
- Competition.

Performance Measures, Cost Management, and Allocation of Resources

Level of Service

Accurate vision statements are important to establishing strategic plans and measuring progress toward achieving the vision. The committee noted that included in the *21st Century Shore Support Infrastructure: Navy Infrastructure Vision and Strategic Plan* is the statement: "Our motto should be: Equal to or better service at equal to or less cost."[2]

This theme was evident throughout the committee's briefings and site visits and has clearly permeated the entire organization. Although this philosophy is typical Navy "can do" and is understandable, it is inconsistent with the goal of reducing the infrastructure to the minimum necessary to satisfy the users' requirements. It establishes a particular mind-set throughout the organization that potentially precludes critical analysis of where the goods or services provided exceed user requirements or where the cost of providing the last *n* percent of quality or quantity is not cost-effective (e.g., as when providing the last 5 percent of the quality or quantity consumes 20 percent of the cost). This kind of critical analysis is a fundamental management tool in determining the "knee of the curve" for providing goods and services and the return on the investment, and it is a necessary component of determining opportunities for reducing the infrastructure without eliminating real required goods and services. If the infrastructure is to be reduced to the level that is necessary, it must be understood and accepted by both the users and the leadership that is acceptable for the infrastructure to provide only the minimum quality and quantity of goods and services to satisfy the users' requirements.

Finding

The current Navy infrastructure vision and strategic plan discussed above may reduce the likelihood of obtaining critical analysis of the value of goods and services provided (e.g., essential services at minimum cost) and thereby impede cost-effective reductions in the infrastructure.

Major Recommendation

The Navy should change its statement of infrastructure vision to "Essential service at minimum cost."

---

[2]Hancock, VADM W.J., USN, Deputy Chief of Naval Operations (Logistics, N4). 1997. *21st Century Shore Support Infrastructure: Navy Infrastructure Vision and Strategic Plan,* Washington, D.C., June 24.

Output Measures

The reason for attempting to reduce and streamline the infrastructure is to reduce costs and free up funds to recapitalize the Navy. To date, streamlining has been done mainly by arbitrarily reducing funds for the infrastructure and charging the infrastructure with developing new and innovative means of providing the same or better service at lower cost. There is, however, no way of knowing if the reductions in funds in different areas are too great or not great enough—i.e., will the reduction in funds result in a loss in capability and readiness, or is there potential for even greater reduction? Without a means of measuring the output of the infrastructure, these questions cannot be answered and the Navy is left with trial and error, which potentially can lead to serious degradation in the necessary support provided to the operating forces.

Traditionally the Navy has, for the most part, used inputs in determining the performance and requirements of the infrastructure, particularly base operating support. For example, the requirement for funding for real property maintenance is often expressed as a percentage of the current plant value (i.e., real property maintenance "should be funded" at 2 percent of the current plant value). The problem with this approach is that it neither identifies the consequences of funding at higher or lower levels nor relates funding to the capabilities of the facilities being maintained to perform the functions required by the user. A second problem is stating a requirement in terms—usually in financial terms (e.g., backlog of maintenance and repair)—that do not describe the consequences or loss in capability if the requirement is not satisfied.

To determine the minimum resource requirements for the infrastructure and relate resource allocations to the ability of the infrastructure to support the needs and readiness of the operating forces, it is necessary to develop measures of performance that capture output. To be most effective, the user(s) must contribute to the process of developing the requirements for output; some such measures for regional bases are discussed in Chapter 2 (in the section titled "Regionalization").

In developing output measures, several steps are required. The process should be iterative and continuous, allowing decision makers, from the program manager to the CNO, to evaluate what is required to support the user and what level of performance should be expected by the user. A recommended process, suggested in Appendix C, assumes that the Navy's regionalization plan is implemented.

Funding Vehicles, Paying for Internal Services, and the Navy Working Capital Fund

Within the Navy and the Marine Corps, many activities provide services to other parts of the organization. When providing these internal services, there is always a concern as to how to allocate resources across the many demands, how to control costs, and how to guarantee that customers' needs are best satisfied. A good internal management process should provide cost visibility, performance measurement, cost containment, and customer control.

There are basically two ways to pay for services provided internally by an organization: customers pay for services as they are provided, or central management gives a budget to the service provider and directs that organization to deliver its services to the internal customers.

Portions of the Navy and the Marine Corps have been using some form of a customer reimbursable system since 1878. As discussed in the Center for Naval Analyses (CNA) report on the Defense Business Operations Fund (DBOF),[3] DBOF was created in 1991 to consolidate all of DOD's reimbursable systems under one set of rules. In 1997, the Office of the Secretary of Defense (OSD) returned oversight of the reimbursable systems to the Services and relabeled them as the Working Capital Fund. Across all of DOD, approximately $75 billion in services are paid for by customers, and the Navy and Marine Corps account for about $20 billion of this business.

Direct payment to service providers is an alternative to customer reimbursements to service providers for services. Mission funding does not provide the same incentives to service providers to reduce costs and improve performance, but it does guarantee some specific amount of revenue to the service providers and thereby allows for a more stable, guaranteed workload.

The Department of the Navy has used both customer reimbursement and mission funding as a means to pay for various infrastructure services for more than 100 years, and is not likely to totally eliminate either process in favor of the other for many more years, if ever.

The committee is aware of (1) many of the advantages and disadvantages of the different means of payment for internal services, and (2) some of the current difficulties that customers have with Navy Working Capital Fund (NWCF) services and/or their costs. This study and this committee, however, were neither designed nor tasked to specifically address NWCF or mission funding issues. The committee's primary purpose in discussing the NWCF activities is to make sure that they are included in future Navy efforts to develop and improve measures of output along with mission funding activities. The NWCF activities are providers of a major function of all infrastructure services, and the exclusion of some or all of them from consideration in developing better output measures for the infrastructure would be a major failing.

Moreover, the committee observes that the NWCF activities are currently under the management cognizance of the financial managers and comptrollers,

---

[3]Trunkey, R. Derek, and Jino Choi. 1996. *The Defense Business Operations Fund (DBOF) Problems and Possible Solutions*, Research Memorandum 95-196, Center for Naval Analyses, Alexandria, Va., March.

and that there is no equivalent of a chief operating officer for this more than $20 billion annual Department of the Navy business. Although the financial managers focus well on the cost inputs for providing NWCF services, these customer-funded services are as lacking in quality output measures as are the mission-funded activities.

Finding

The Navy does not have sufficient output measures to properly determine the value of and minimum essential needs for infrastructure goods and services, and to justify the allocation of resources to provide such infrastructure goods and services.

Major Recommendation

The Navy should establish a management information system to track support-system performance and costs. This system should be based on an integrated set of cost and performance metrics that are developed using fleet user inputs as well as those of service providers.

Recommendations

- *The CNO should put a process in place across the Navy, in each major claimancy and in each region, to develop and maintain output measures for each major function or activity that is provided by the infrastructure. A recommended process for accomplishing this is outlined in Appendix C.*
- *The CNO should ensure that users have a prominent role in developing the requirements for the goods and services to be provided and the funding required to provide them. Users should focus on essential, cost-efficient support rather than better support and on acceptable risk rather than risk avoidance.*
- *N4 should initiate a fresh look at current Navy metrics, such as the requirement for real property maintenance (RPM) and the backlog of maintenance and repair, by reviewing best business practices.*
- *There should be two sets of metrics, one top-level and the other management-level. The top-level metrics should reflect the overall goals of the infrastructure and should measure system cost and effectiveness at the major claimant and Navy-wide level. The management-level metrics should be more clearly related to day-to-day resource decisions at the region or activity level. The metrics should be constructed so that management-level process metrics can be aligned with top-level strategic metrics.*
- *The CNO should direct that N8 set up a system that tracks the top-level infrastructure metrics and helps to assess cost-effectiveness in meeting fleet needs. The elements of such a system would include (1) using a cost model such*

as the COBRA model[4] (used to establish a cost baseline for comparison of comparable mission and base functions across DOD for the base realignment and closure process in 1991, 1993, and 1995) to establish a cost and performance baseline for infrastructure activities; (2) ensuring that the full cost to the Navy of military manpower is included as a cost variable; (3) linking essential fleet requirements to infrastructure costs; and (4) taking full account of the feedback effects of changes in the support system as a consequence of changes in the operating fleet.

- *The CNO should request that the Secretary of the Navy (SECNAV) task the Assistant Secretary of the Navy/Financial Management and Comptroller (ASN/FM) to appoint an NWCF chief operating officer, responsible for (1) providing greater visibility regarding the cost of individual NWCF activities, including related overhead costs, and (2) overseeing the development of output measures for NWCF activities, to include giving users of services (major claimants) the opportunity to participate in the development of such metrics.*

Finding

The Navy does not have a mechanism or process for determining, on a continuing basis, current industry best practices and performance for comparable functions.

Recommendation

The Navy should invest in a resource person for each regional commander's and major claimant's staff who is familiar with, and has access to, industry and trade associations and consultants who can provide up-to-date information on current industry best practices for each of the infrastructure functions provided by the region. This information should be used in setting cost and performance goals. A network of such persons with expertise in best practices should be developed for sharing the practices and lessons learned in applying them within the Navy.

---

[4]R&K Engineering Planning Systems. 1997-1998. *Cost of Base Realignment Actions (COBRA)*, Roanoke, Va. Available online at <http://www.rkeng.com/projects/cobra.html>; U.S. General Accounting Office. 1997. *Military Bases: Lessons Learned from Prior Base Closure Rounds*, GAO/NSIAD-97-151, Washington, D.C., July; U.S. General Accounting Office. 1995. *Military Bases: Analysis of DOD's 1995 Process and Recommendations for Closure and Realignment*, GAO/NSIAD-95-133, Washington, D.C., April; U.S. General Accounting Office. 1993. *Military Bases: Analysis of DOD's Recommendations and Selection Process for Closures and Realignments*, GAO/NSIAD-93-173, Washington, D.C., April.

Information Technology for Infrastructure Management

At the start of the information age, the DOD led the way in developing both computers and networks. It dominated the marketplace, funding the research and setting the priorities. In 1955, the Navy was considered to have the best computer systems in the world. But a sea-state change has occurred, and today's information technology (IT) industry is leading the charge. DOD currently represents less than 5 percent of IT business. The objectives are to improve connectivity, speed of information flow and information sharing, streamlined operations, and new business models. Success derives from getting the right information to the right people in the right time frame. Connectivity, content, cost, control, sharing, and collaboration are the essential elements to success.

A Navy-wide Information Plan

The committee acknowledges the difficulty and enormous costs associated with creating an enterprise-wide information plan for an institution as complex as the Navy. However, the payoff is potentially huge because information technology can have an impact on the infrastructure and on shore readiness in two ways. The first is to increase capabilities and provide services in ways that were not possible before. The second is to reduce the costs of providing the fleet with current shore-based outputs by using new means to provide the services. In either case, information systems rarely result in increased capabilities or reduced costs simply by automating tasks being performed by personnel. Rather, information systems are a lever for increasing capabilities or reducing costs by enhancing access to and accountability for data and services.

The industrial age revolutionized the Navy by giving individuals control over thousands rather than tens of horsepower. This magnification of human labor revolutionized society as well. A modern revolution in information technology offers the Navy another opportunity to make startling improvements. IT provides individuals with unparalleled control over goods, services, and activities that cross the barriers of time and distance. Exploiting this newfound power is key to a cost-effective shore establishment. Information technology in the 21st century is the lubricant that will allow the Navy business to flow with less friction and waste. It holds the potential for integrating ship and shore in ways that were never before possible.

Given at the end of this section are the committee's recommendations for evolving a Navy-wide information plan with the following attributes:

- *Navy-wide information space (infospace):* defining, creating, and maintaining it;
- *Service access:* the key to consolidation of infospace-based Navy shore services; and

- *Decision support:* leveraging the power of managers for cost-effective decision making.

Navy-wide Information Space and Its Components

The concept of a Navy-wide information space (infospace) is that of an adaptive system defined at any time by a set of performance standards for timely and effective information delivery throughout the Department of the Navy. The infospace includes basic components such as an information network infrastructure and information workstation capabilities that can be adapted to IT technology advances; associated training; and technology refresh cycles. However, it is emphasized that the infospace is a system characterized by a set of connectivity and service goals rather than a set of hardware and software goals. The particular technology solutions can vary, but the goals can and should remain relatively constant. Establishment of connectivity and service goals creates a benchmark against which progress can be measured without constraining the combination of solutions applied to the problem. Although this discussion is aimed primarily at the Navy, the infospace will eventually have to embrace all of DOD. Many of the prescriptions in the following sections can be applied DOD-wide.

The purpose of the infospace is to provide an infrastructure for delivery of information and services from any provider to anyone in need of a service. This approach must use Internet technology, rather than attempt to provide new point-to-point connections for each client-server combination. The concept of major information services requiring specific connectivity and specific workstations is no longer economically feasible. There must be a standard vehicle for delivery of and access to information and services. It should include not only a network with sufficient connectivity and bandwidth but also a standard information workstation that all service providers can depend on and develop around (i.e., a standard for developers and providers of software applications).

The desired infospace is, therefore, a trusted capability that all developers can depend on (e.g., as it is possible to depend on the existing telephone system). In an ideal system, the issues of how to get connected to users, where those users are, and what capabilities they have will disappear. Such a trusted service delivery system is critical to the Navy's consolidation and regionalization cost-cutting efforts. If all service providers know that they can reliably deliver their service to any user anywhere in the Navy, the design and development task is greatly simplified. In addition, issues of geography and distance can be subordinated to those of cost and effective service. If commanders and managers can depend on the infospace to access and obtain service in a reliable and timely manner, the management need for "controlling our own" is diminished. This lowers many of the cultural barriers to consolidation and regionalization. The key is that on both the provider and the user sides, the infospace must be trusted in terms of security, access, timeliness, and usability.

There are four basic components to a Navy-wide infospace, each requiring standards:

1. Network infrastructure,
2. Standard information workstation,
3. Training, and
4. Refresh and replacement cycle.

Each of these components works with the others to create the dependable delivery of information services across the Navy. Note that these components are a set of standards around which a variety of capabilities can be developed. The important thing is to select the standards and begin building immediately. These will serve to focus the now varied and disjointed efforts. The goal is not to stop the many organizations within the Navy from developing the capabilities that they need to get their jobs done. The goal is to provide a unified vision so that the efforts of all can provide a cost-effective solution for all. The challenges faced in the development of an infospace vision cannot be used as an excuse to further slow the deployment of Internet and computing technology throughout the Navy.

Network Infrastructure. The key to the infospace is the establishment of Navy-wide secure and unsecure (broadly accessible) internets, the communications backbone that ties the structure together. Again, the infospace network, however, is a set of goals and standards for connectivity, security, and service, not a set of technology requirements. There must be a continual process of evaluation of the performance of the infospace relative to goals, and evaluation of the goals themselves. These goals provide the focus for technology acquisition. Programs that promise networked voice and video services while the Navy cannot yet guarantee universal Web access are counterproductive. The key goal areas are connectivity, timely service standards (bandwidth), electronic mail (e-mail), access to the World Wide Web, and security.

Connectivity. The key goal is that every standard information workstation in the Navy will be electronically connected to every other information workstation and to every information service. Note that technological connectivity is different from authorized connectivity. Whether an information workstation can access a given service should never be a question of stringing wires or of acquiring hardware. It should be only a negotiation of authorization. The question should never be, Is it possible to get timely access? but rather, Should access be authorized? The infospace technology should never be a barrier; it should always be a facilitator. Barriers, when needed, should be created intentionally. For example, barriers to data communication should exist only for purposes of good management and security.

A Navy-wide connectivity scorecard should be created as a key management tool. The scorecard will continuously monitor the percentage of information workstations in the Navy that meet the connectivity goals. This information will provide a clear focus for the management and development of the infospace.

Timely Service Standards (Bandwidth). A prime reason for lack of trust in information technology services is poor response time. A service that is too slow will not be used. Users will find workarounds that will avoid the infospace service. These workarounds will drive up costs in ways that are undesirable and hard to manage. Infospace service providers should not be allowed to deliver untimely services. Ineffective, globally provided services will encourage local managers to build up their own individual capabilities and thus will defeat the desired cost savings many times over.

Minimal bandwidth standards for each information workstation should be defined. This means not only technical communications capacity to the end station but also total network throughput. A minimal standard might be that "a 10K Web page can be delivered from any server in the Navy to any information workstation in the Navy in an average of 10 seconds or less."

Delivery of a service in 10 seconds is not actually acceptable for interactive work, but it does guarantee some access to everything in the Navy. Another helpful standard might be that "a 10K Web page can be delivered from any server to its primary users in an average of 2 seconds and a maximum of 10 seconds."

A variety of standards can be set that define terms of actual services to be delivered, not technology to be purchased. The key is timely service, whether ATM, 10BaseT, switched Ethernet, or plain telephone lines are used.

There should be a regular system for monitoring whether or not each separate goal (standard for timely service) is being achieved. In the case of WWW pages, COTS proxy servers can provide all of the necessary statistics to easily monitor key performance. Periodic testing and audits of performance can be carried out. Because these goals are always stated in terms of performance perceived by end users, any users who feel that they are not being served can run their own test with a wrist watch. The scorecard for compliance is easily understood and easily verified. Compliance with timeliness goals cannot be overemphasized. Poor service is useless service, and users will find other more expensive ways to do their job.

Electronic Mail. Electronic mail (e-mail) is the premier collaborative tool in the Internet age. All information workers in the Navy should have an e-mail address that can follow them in a robust fashion as they move through their careers. It should be possible to rapidly update such addresses with minimal effort in response to job changes. It should also be easy to locate and separate the e-mail address of an individual job function (e.g., N4) from that of its current occupant (e.g., the N4). Such knowledge, of course, is subject to authorization. Such "white pages" services can be delivered via the WWW and thus exploit the same authorization structure.

A key to rapidly organizing teams to work on problems when they are not co-located is the construction of a team mailing list. Every officer with administrative responsibility should know (either personally or through trained staff) how to create such a mailing list in less than a day. This capability provides for rapid organization of personnel without the restrictions of time and distance.

Access to the World Wide Web. The World Wide Web (WWW) is the primary vehicle for universal access to infospace services. Web servers are easy to set up and easy to maintain. Training for creating and managing Web sites should be widespread among Navy personnel. Information made available on the WWW should replace most hard-copy manuals and general information materials. In specifying the information workstation, a well-defined set of plug-ins or help applications should be included. Among this set would be office productivity tools and a standard page-publishing tool such as Adobe's portable document format (PDF)/Acrobat. These standard applications help provide for almost instantaneous delivery of manuals, procedures, and training materials. If every information workstation has infospace connectivity and standard WWW browsers and help tools, then electronic publication of materials can become a reality. Instead of archiving, shipping, storing, and updating tons of paper publications, they all can be distributed and accessed on an as-needed basis. This can result in a huge savings and provide more timely and accurate information.

It is important to recognize that the Hypertext Mark-up Language/Hypertext Transfer Protocol (HTML/HTTP) standards together form the backbone of the WWW. With these, a vast amount of information and service can be shared from Navy databases. HTML alone can provide many of the access and information needs required to support consolidated or regionalized shore services. In addition, standard Java plug-ins can allow consolidated providers to download specialized interfaces to their services.

Security. The military's need for secure communications is obvious and pervasive throughout the infospace. Systems must be in place to monitor compliance with security standards to create a trusted environment.

One aspect of the security architecture that receives less attention than communications and computer security is authorization for services. A uniform authorization architecture is needed. When individuals create a new service to be delivered across the infospace, it must be relatively straightforward for them to define which job functions are authorized to have access to and make requests for that service. They must be able to easily accept requests for service and validate those requests against the authorizations defined for the service. It is not acceptable for every new service to develop its own mechanism for determining who is trying to access the service and whether or not they have permission. This is part of the infrastructure that must be generalized to permit free but appropriate access across the infospace. Needed are standard mechanisms for granting permission as well as standards for validating digital signatures. Not all of these needs are currently satisfied by COTS products.

Standard Information Workstation. Every information worker in the Navy should have access to a standard information workstation, which is key to a successful Navy-wide infospace—a network connection between service provider and user is not by itself sufficient. If the service provider cannot depend on baseline capabilities at the user end, delivery of the service will not be successful or will be much more complicated than necessary. Information workstations may or may not be the same as warfighting computers. The key is that there be a standard set of capabilities that service providers can test their software against and then depend on. Users need to know that if they have a standard information workstation, they will have access to any of the services offered over the Navy-wide infospace system. Standardizing the information workstation capabilities also simplifies the training process for users.

The information workstation should not be a vendor standard; it should be a standard set of testable capabilities. Where there is a vendor-specific standard, the opportunity for competition is reduced and the value of the standard information workstation declines in the long run. In terms of hardware, vendor-independent standards are quite straightforward. Vendor-independent software standards are much more difficult. In many cases, vendor-specific software standards are unavoidable. Where possible, two alternatives should be allowed. This complicates the standard somewhat but preserves the critical competitive environment. However, any alternative must be a viable competitor in the open marketplace rather than one supported by its privileged position as a Navy standard.

The standard information workstation should include:

- A standard operating system and "windowing" interface system;
- Standard office tools such as word processing, spreadsheet, and presentation software;
- WWW browser plus standard help applications; and
- A downloadable software translator standard such as Java.

A standard operating system and "windowing" interface system can simplify training and ensure users' confidence in their ability to access the infospace services. With standardized office tools, the training is again simplified. In addition, the ability of personnel to collaborate across the Navy-wide system is greatly enhanced. If everyone has access to the same word processor or spreadsheet, then sharing documents or budgets via e-mail can become a reliable and trusted form of communication. If these tools are not interoperable, then trying to share information becomes a frustrating experience.

The WWW browser standard is the tool for universal access to consolidated infospace services and defines a unified window on the world. A downloadable software standard allows providers to specialize the access to their services. It is very important that this standard be carefully constrained and monitored. A selected standard must have provable security properties to prevent viruses and other threats, and it must be reliable. Minor variations in the downloadable software standard can cause frustrating user interface failures.

Training. The purpose of creating the infospace is to facilitate access to information and services. Technology alone will not accomplish this. It is essential that training programs be put in place to ensure that everyone knows how to access infospace services. Generational issues will arise. Some personnel over age 50 will never use a computer as part of their everyday activity. Trained staff must support such people. Most admirals do not type their own letters, and some do not lay hands on a computer, provided that all support staff are appropriately trained. For all newly trained personnel, both enlisted and officers, basic access and navigation skills should be required training. It is of little value for a regionalized service to provide infospace access when few of its customers have the training necessary to exploit such access. Again, a dependable and predictably trained base of users is essential to the delivery of infospace-based services.

Refresh and Replacement Cycle. The cardinal rule of computing technology is that it will change. Any model for computing that is based on "buy once, use forever" is foolish—the Navy infospace must be an adaptive system. End user computing technology should be amortized over 4 years, with slightly longer periods for network services. There must be a plan for continual upgrade and replacement. The power and capabilities of computers are doubling every 18 months. This means that a 4-year-old machine is five times slower than a new machine at the same price. A problem with Navy-wide infrastructure standards is that they cannot change willy-nilly. Everyone must depend on the standard. The nature of computing, however, demands responsiveness to continual change.

If a 4-year rolling plan is adopted for upgrades, the standards can also be placed on such a plan. For a given year, one can stipulate that 4 years from now, every information workstation in the Navy will have capability *X*. Setting a new standard defines the upgrade. When service providers want to utilize the new capability, they can request that particular users upgrade sooner or plan a deployment that is in line with the upgrade plans.

The technology refresh plan should also provide for the discarding or upgrading of non-compliant and unused applications. When yearly standards are set, service developers are put on notice that within 1 year, 25 percent of the information workstations or Internet capability will no longer support their service and that within 4 years none of the information workstations will. Such a planning cycle gives clear direction to the upgrade process that everyone can work within.

The technology refresh plan also provides for upgrades to training. As the technology requirements within the infospace are changed, the corresponding training can be changed. With the new capabilities will come a new generation of personnel prepared to exploit those capabilities.

Service Access

The key to consolidation of infospace-based Navy shore services is to provide access to those services that is as good as or better than before. A Navy-wide infospace can facilitate such savings. A centralized automobile fleet that provides reservations over the Internet is more accessible than several with a few vehicles scattered at multiple sites. The goal is to provide access to services via the infospace rather than physically locating the services near the customers. This effectively moves the service from "down the street" to "on the desk." Such electronically available shore services will be more tightly integrated and accessible to shipboard personnel than any number of physical "storefronts."

Each time a service is consolidated to a centralized location from a number of independent entities, a new infospace application should be created to provide access to the consolidated service. This provides service users with more effective and timely access than before, while at the same time reducing the costs of supporting the service. In order to meet such a goal, new strategies for developing information technology applications must be adopted. The old-fashioned COBOL-oriented multiyear development methods are completely unworkable in this domain. The very large "do everything but the dishes" software contracts will not serve this need. Software must be created and changed rapidly to provide exactly the service information required. Changes brought on by vast amounts of cheap computing power, universal networking, and standardized workstations can make such developments possible. The development time for new applications for small organizations must be cut from years to months (3 to 5). Otherwise, the costs of supporting large software installations will remove any cost savings that may result from the consolidation. Old-style software methods frequently transferred costs from service providers to IT providers without any real cost savings. This cannot be repeated.

Service Expectations. A key to effective network-based rather than location-based services is a set of uniform expectations about such services. One of the major costs in IT deployment is the training of users. The Navy cannot afford to create a major training program for every new service application. To eliminate the training burden, all services within the infospace should support the following features in a standard way.

- Provide uniform mechanisms for finding desired services.
- Supply sufficient information for intelligent selection of services by users where more than one alternative is available.
- Implement uniform mechanisms for requesting services and in particular for requesting authorization to receive service.
- Provide clear, uniform tools whereby users can estimate the costs of a requested service. This will promote effective shopping for services. Users cannot be expected to make cost-effective use of resources if the cost implications of requesting a service are not clear.
- Provide users with accurate information about the status of their request.
- Provide users with accurate metrics of overall performance (i.e., average time to delivery, average number of faults, and so on). Simply making such performance information known to users will improve cost-effective delivery of service.

When human-based services are replaced with networked information-based access to services, trust must be established. Current Navy personnel have clear expectations of the control that geographic proximity and command structure can provide. Effective officers and other managers have learned how to get results from such services. When access to such services is through an infospace, new expectations must be established. Providing a uniform set of such expectations, which apply to all services obtained over a Navy-wide infospace, will sharply lower the barriers to replacing personal access with electronic access.

Metrics for Infospace Service Success. Information technology has a mixed record in the area of cost reduction. With each such project, metrics for success should be clearly established up front. A key metric in this area is staffing impact. Every information technology application should be able to clearly state, "If this IT application is put into place, then the total Navy staffing will be reduced by *X* number of people within *Y* years."

Having established such a claim, the IT project should be held to it. In particular, the hidden staffing requirements must be made known, including IT support people and user area staff required to use the application. Replacing user staff with IT support staff will generally increase costs rather than reduce them. It must be clearly shown that the IT staff is significantly smaller than the staff that the application is replacing. It must also be shown that the IT project is not pushing work out of the centralized area into the user areas. This is frequently done when centralized impact is easy to see while user area work is considered to "blend into the overhead." Using IT to move work does not save money. IT saves money only when it eliminates work entirely.

Decision Support

The payoff from these information technology investments will be major improvements in the ways functions are performed and decisions are made within the activities composing the infrastructure. When IT is used to leverage the power of decision makers to make cost-effective choices, then costs can be reduced. For example, alternative ways to deploy helicopter assets in response to humanitarian requests could be examined, with both costs and time-line parameters displayed; or information service that quickly displays all of the choices with their costs could enable in-transit personnel to select cheaper quarters.

Information technology can rapidly generate and present alternatives in a way that is not possible otherwise. Information systems can also clarify the consequences of decisions. A simple example is the ship berthing and movement tool created by California Polytechnic State University and used in the San Diego region. This tool allows harbor personnel to rapidly work through various ship movement plans to determine which might meet the goals with less cost in time and money. It is cheaper to move ships on a screen than under real power. The Navy needs more of these kinds of decision support tools that clarify cost consequences of decisions.

The strategy for new applications of software and databases will require professional attention. Historically, the information systems organization in major companies was responsible for applications running on mainframe computers, with users operating dumb terminals. The widespread availability of personal computers did little to enhance the utility of the functional stovepipes of information that have evolved within the departments, such as sales, finance, and customer service, and provided little improvement in productivity overall.

In the 1990s, client-server-based systems evolved as the model for distributed computing. However, interdepartmental data analysis and management required enterprise resource planning (ERP) tools involving specialized analysts to tailor the interfaces for each client's request. Today, as the World Wide Web, browsers, and other tools mature, the architecture for enterprise-wide data access and analysis can evolve smoothly from a hardware, software, and configuration management perspective. Thus, the committee believes that both the client-server and Web-based architectures have their own advantages; the Navy will require professional assistance in managing its implementation program.

Major Recommendation

The Navy should define and implement the concept of a Department of the Navy-wide information space (infospace) using a set of standards—some of which are now being partially implemented by the Information Technology for the 21st Century (IT-21) initiative—to serve both shore and afloat activities. The infospace should be defined by performance standards for timely and effec-

tive delivery of information and services throughout the Department of the Navy, including the fleet and its support. It should include a network infrastructure and a standard information workstation for every responsible information worker in the Navy, not just those in the operating forces. The infospace should be used as the primary vehicle for delivery and integration of information about the entire Navy infrastructure to include the following functions: acquisition, central logistics, central personnel, communications, force management, installations, medical functions, quality of life, science and technology, and training. Funding for this effort, including funding for essential technology upgrades and related training, should be identified and protected within the Department of the Navy.

Recommendations

- *The Chief Information Officer (CIO) should:*

— Establish performance, service, and communications standards for the infospace that are not adequately addressed in the current draft guidance (Information Technology Standards Guidance).[5] *These standards must be performance based and not be a set of technology requirements.*

— Advise the N6 on which Navy-wide site licenses should be negotiated for use on standard information workstations. Such licenses will encourage interoperability, intercommunication, and standardization.

— Ensure that there is a Navy information management system both to monitor IT performance and cost and to track metrics for successful delivery of services within the infospace.

— Act as a facilitator and arbitrator for cross-stovepipe IT applications.

- *The CNO should require basic infospace access and navigation training for all new enlisted and officer personnel, and he should address the IT specialist personnel training and retention problems caused by commercial market forces. Possible alternatives are using educational subsidies with extended tours to train and retain IT specialist personnel, and outsourcing IT specialty needs.*
- *All Navy managers should:*

— Ensure that IT applications derive from business change initiatives;

— Use the Navy-wide infospace to create paperless business transactions;

— Use decision support tools and performance metrics to elucidate cost and performance consequences of business decisions; and

— Encourage use of the Navy-wide infospace for disseminating good IT practices.

---

[5]ITSG Integrated Product Team. 1998. *Information Technology Standards Guidance, Version 98-1.1,* Chief Information Officer, Department of the Navy, Washington, D.C., June 15. Available online at <http://www.doncio.navy.mil/itsgpublic>.

Competition[6]

Recent experience has indicated that substantial savings can result from competition.[7,8] From 1978 to 1994, the DOD held more than 2,000 public-private competitions with an average planned savings of 30 percent. These planned savings were seen in all the military branches and for virtually all types of commercial activities. Based on these experiences, in 1995 the Navy planned to outsource about 25,000 positions, including 3,400 military billets, and reflected these savings in its out-year plans. Although the number of competitions held through February 1998 was for only 2,100 positions, the Navy increased the goal of completing A-76 competitions, involving 80,500 positions, including 10,000 military billets, by the year 2002.[9,10] In light of past experiences, many have questioned the Navy's ability to realize savings of these magnitudes in the future. Here, evidence is reviewed and ways are proposed to improve the success of the program.

Many excuses are given for failing to realize the projected savings, but they can be grouped into three categories: (1) the number of positions available falls far short of the required target because the function performed is deemed inappropriate for competitive sourcing, (2) the number of positions actually competed is below expectation because local constraints usually govern the process, and (3) the realized savings percentage is lower than forecasted because previous actions already had captured the largest source of efficiencies or the budget already reflected the cost of the most efficient organization (MEO).

A review of data and trends indicates that more savings are not being realized through competitive sourcing because most jobs ashore have not been competed. Specifically, according to the Navy's records, there are 300,000 commercial activity positions (60 percent of which are civilian) and, at most, only 10 percent have been previously competed. In addition, many military personnel remain in commercial activities, but personnel policies limit the number of bil-

---

[6]Privatization is often cited as a preferred method of improving performance or reducing cost. However, the committee believes that competition tied to a particular service and marketplace is the key enabler that stimulates innovation and improvement in both government and private activities.

[7]Marcus, Alan J. 1993. *Analysis of the Navy's Commercial Activities Program,* CRM-92-226, Center for Naval Analyses, Alexandria, Va., July.

[8]Tighe, Carla E., Samuel D. Kleinman, James M. Jondrow, and R. Derek Trunkey. 1996. *Outsourcing and Competition: Lessons Learned from DOD Commercial Activities Programs,* CNA Occasional Paper 23, Center for Naval Analyses, Alexandria, Va., October.

[9]U.S. General Accounting Office. 1998. *Defense Outsourcing: Impact on Naval Sea-Shore Rotations,* GAO/NSIAD-98-107, Washington, D.C., April.

[10]Trunkey, R. Derek, Benjamin P. Scafidi, Francis P. Clarke, Cheryl Kandaras, Andrew M. Seamans, LCDR Carolyn M. Kresek, USN, Robert P. Trost, Angela L. King, Christine H. Baxter, Kerensa E. Riordan, Steven Smith, and Michael Ye. 1998. *Moving Forward with A-76 in the Navy,* CRM98-9.10, Center for Naval Analyses, Alexandria, Va., April.

lets available for competition. A recent study showed that, in the Navy, only 16,800 of 138,000 military shore billets are available for competitions after accounting for policies such as sea-shore rotation and homebasing.[11] While this should be sufficient to complete the current rounds of competitions, such large exemptions limit the Navy's flexibility in targeting activities and consolidating them for larger competitions. Moreover, in past years, 34 percent of Navy competitions, many of them very large, were canceled. By comparison, only 3 percent at the General Services Administration were canceled because senior leadership supported the program and demanded that it be executed. As mentioned above, the Navy plans to announce competitive sourcing for 80,500 positions over a 5-year period and is building its candidate list from claimant inputs. In 1997, the Navy announced the proposed competition of 10,000 positions. The goal for 1998 was to announce 15,000 positions, but the Navy was unable to achieve that. The Navy has so far completed competitions for only 250 positions out of its goal of 80,500. Clearly the cause for these shortfalls is a major concern because government personnel represent a large fraction of infrastructure costs.

Sea-Shore Rotation Policy Constraints

Because the policy of reserving shore billets for military personnel after tours at sea places a major constraint on the number of jobs eligible for competition, the committee examined the issue in some depth. The Navy needs to attract and retain quality sailors. A component of the overall personnel management process is the rotation of the career enlisted force between sea and shore billets. The purpose is to reduce the time deployed and the associated stress of operational jobs and separation from family. Another reason is to enhance skills for later deployments. The Navy fears that more sea time will hurt both recruitment and retention.

The rotation policy is executed by reserving many shore jobs for military personnel. In recent years, the sea-shore rotation objectives have been coming more and more into conflict with Navy objectives to reduce the number of personnel providing support ashore. A recent homebasing policy to rotate personnel into shore billets near their previous sea assignment introduces additional constraints. If cutting military personnel ashore is required to improve efficiency and if that will hurt personnel readiness, the Navy has to consider alternative ways to soften the impact and make a Navy career attractive. This section

[11]Trunkey, R. Derek, Benjamin P. Scafidi, Francis P. Clark, Cheryl Kandaras, Andrew M. Seamans, LCDR Carolyn M. Kresek, USN, Robert P. Trost, Angela L. King, Christine H. Baxter, Kerensa E. Riordan, Steven Smith, and Michael Ye. 1998. *Moving Forward with A-76 in the Navy*, Research Memorandum 98-9.10, Center for Naval Analyses, Alexandria, Va., May.

reviews some of the issues and considers ways to satisfy both personnel and efficiency objectives.

Measuring Rotation

The Navy classifies all its billets with a sea-shore code. Sea duty billets are those in deployable units or at some of the overseas ashore installations. Student billets are classified separately and are not included in the computed sea-shore numbers. The metric used by the Navy to monitor sea-shore rotation is the sea-shore ratio. This indicator is the simple ratio of time in sea duty to time in shore duty. It is computed only for E5 to E9 billets, because these are considered the career pay grades. Overall, the Navy calculates its sea-shore ratio at about 3.8:3. That is, a career sailor is expected to spend 3.8 years at sea for every 3 years ashore. There is considerable variance across ratings. An important attribute of the sea-shore ratio is that the DOD and the Congress generally accept it as a tool to manage the Navy career force.

There are several concerns with using the sea-shore ratio as a management tool:

- The ratio is for billets and not personnel. Although manning practices vary, one should protect only military shore billets that correspond to sea billets that are actually filled by people in pay grades E5 to E9.
- The metric covers all E5 billets. Although some of those billets are filled by E5s in their first term, rotation is designed for careerists. By including only positions traditionally filled by E5 careerists, the process could have a significant effect on the ratings with 6-year initial obligations.
- Student billets ashore are excluded from the shore part of the ratio, although one of the reasons to return personnel to shore is to develop skills that will increase productivity at sea.

The committee examined sea-shore ratios for personnel instead of billets, using the Navy's classification of sea and shore billets. Figure 3.1 displays these ratios over time, with the following observations. The jump in 1989 reflects a change in the definition. The top curve represents the ratio for personnel using the Navy's practice of including all E5s and above. That curve shows that the ratio for personnel has averaged 3.3:3 over the past few years. That is, when personnel rather than billets are looked at, there appears to be more than a 10 percent drop in the ratio (from 3.8:3 to 3.3:3). The lower curve includes only career personnel. Comparing the curves suggests two distortions caused by including first-term E5s. It makes the sea part of the career appear larger and the long-term decline look more dramatic than it was. Overall, actual Navy sea-shore ratios for careerists are less than 3:3. Individual ratings may still have

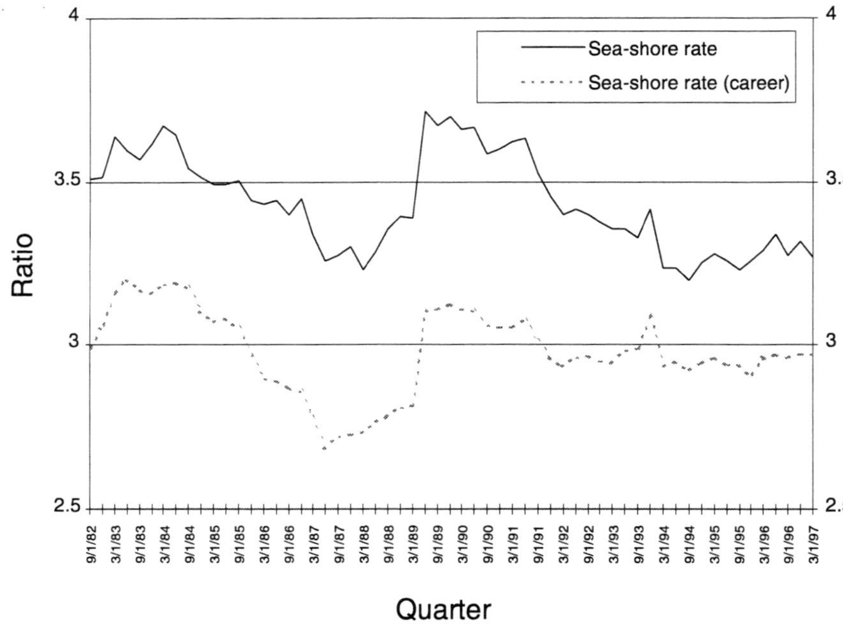

FIGURE 3.1 Sea-shore ratios for personnel, 1982-1997.

considerably higher ratios. The actual sea-shore ratio has been relatively flat since 1991 and generally declined during the drawdown.

Effects of Sea Duty

The principal reason for managing sea-shore rotation is to keep extended sea duty from reducing retention. In surveys, personnel frequently report family separation as a key reason for leaving the Navy. However, analysis of actual retention rates shows only modest impact of sea duty and time underway on retention. An extensive study in the 1970s showed that a 10 percent increase in expected sea duty (e.g., from 50 to 55 percent) would decrease retention by 3 percent (from 25 to 24.1 percent).[12] A later study of retention in the 1980s found similar effects for measures of personnel tempo (perstempo).[13] This study

[12]Warner, John, and Matthew Goldberg. 1984. "The Influence of Nonpecuniary Factors on Labor Supply: The Case of Navy Enlisted Personnel," *Review of Economics and Statistics,* February, pp. 26-35.

[13]Cooke, Timothy, Alan Marcus, and Aline Quester. 1992. *Personnel Tempo of Operations and Navy Enlisted Retention,* Research Memorandum 91-150, Center for Naval Analyses, Alexandria, Va., February.

looked at actual deployment lengths, time between deployments, and time underway when not deployed. Perstempo had the most deleterious effect on the retention of married personnel. But even here, the greatest impact was from very long deployments, those of more than 8 months. The retention declined from 40 percent to 30.5 percent. Other effects were more modest and in line with the earlier study.

The above studies also found that reenlistment bonuses can compensate for additional sea duty. In most cases, reenlistment bonuses that are equivalent to less than 10 percent of regular military compensation (base pay and tax-adjusted allowances) would offset as much as a 25 percent increase in perstempo.[14,15] Married personnel would require higher bonuses. Some would still leave, but others originally leaving for different reasons would later reenlist and offset the loss.

Hidden Costs of Rotation

Fairly recent research[16-19] has uncovered a hidden cost of reserving shore billets for military personnel. Activities with the highest percentage of military personnel produce the greatest savings when competed, i.e., opened to competitive bidding. Controlling for the size of the competition and the type of activity competed, public-private competitions of military positions produce 50 percent more savings than competitions of civilian positions. This strongly suggests that the Navy pays dearly by retaining a military person unnecessarily in a shore position that could be competed. Although these results may have been unexpected, they are consistent with other general observations:

- Young individuals tend to be less productive than older individuals, at

---

[14]Cooke, Timothy, Alan Marcus, and Aline Quester. 1992. *Personnel Tempo of Operations and Navy Enlisted Retention,* Research Memorandum 91-150, Center for Naval Analyses, Alexandria, Va., February.

[15]Sharma, Ravi, and Henry Griffis. 1995. *Implications of Changes in Time Spent at Sea,* Research Memorandum 94-150, Center for Naval Analyses, Alexandria, Va., April.

[16]Marcus, Alan J. 1993. *Analysis of the Navy's Commerical Activities Programs,* Memorandum 92-226.10, Center for Naval Analyses, Alexandria, Va., July.

[17]King, Angela L., Angela M. Rademacher, and R. Derek Trunkey. 1996. *An Examination of the DoD Commercial Activities Inventory Data,* Information Memorandum 471, Center for Naval Analyses, Alexandria, Va., December.

[18]Trunkey, R. Derek, Robert P. Trost, and Christopher M. Snyder. 1996. *Analysis of DoD's Commercial Activities Programs,* Research Memorandum 96-63, Center for Naval Analyses, Alexandria, Va., December.

[19]Snyder, Christopher M., Robert P. Trost, and R. Derek Trunkey. 1998. *Bidding Behavior in DoD's Commerical Activities Competitions,* Research Memorandum 97-68, Center for Naval Analyses, Alexandria, Va., January.

least between the late teens and mid-fifties. Contractors often use older workers to do work previously done by younger military personnel.

- Turnover reduces the productivity of workers. Rotation by definition moves people continuously through the activity. With a 3-year rotation, the average worker has only 18 months at the activity. Contractors keep workers longer and lessons learned are more likely to persist over time.
- Organizations that are predominantly military have limited ability to use part-timers and temporary workers. Contractors can, at apparently less cost, use overtime during surge workloads and reduce the number of full-time employees.
- Military personnel are not accounted for in local budgets, because these costs are handled centrally. Although local activities do not purposely overuse or misuse personnel, costs are hidden in a budget sense, and no one can free up funds for other uses by cutting military personnel. As a result, the system is biased toward having excess people.
- The demand for work performed by military personnel is often unconstrained by cost. For example, civilian depot costs are often covered by reimbursements from the fleet, whereas the military intermediate maintenance facility costs are not. When there are choices, the fleet tends to use the latter facilities. Although depots attempt to bring down their costs, the intermediate maintenance facilities have no such pressures.

Reconciling Efficiency and Personnel Objectives

Some shore jobs must remain military. Some of these jobs require personnel with recent operational experience or provide additional skills for later tours. But for many jobs, this is not the case, and in most cases there is a hidden cost.

The Navy faces a dilemma in deciding which shore billets can most easily be competed or staffed by civilians. If it eliminates general billets that are unrelated to sailors' jobs at sea, it has reduced its flexibility in assigning personnel ashore. Training and maintenance billets, for example, require personnel with the appropriate rating and Navy enlisted classification (NEC). Yet the rotation needs of individual ratings may shift over time and the shore billets may not match. The clear point here is that the Navy must consider the efficiency gains, the impact on skills, and the flexibility in making assignments when it considers which functions and billets to target for competitive sourcing and other infrastructure efficiencies.

As a general observation, the committee believes that the process for selecting military billets for competition and other efficiencies is weighted too heavily to standard sea-shore goals. For example, the Deputy Chief of Naval Operations, Manpower and Personnel (N1) currently has veto authority over the inclusion of military shore billets. It is the committee's view that the decision needs to be based on the inputs of all stakeholders. Disagreements could be resolved by the Vice Chief of Naval Operations (VCNO) as the final arbitrator.

There are many ways to reconcile efficiency and personnel concerns. For one, the Navy should try to identify alternative ways to improve sailors' quality of life. For example, many sailors find workloads while in home port to be particularly onerous. Since about half of the time in a sea billet is spent at the home port, those working conditions should be reviewed. Tiger teams on the base could move from ship to ship to reduce long workdays when the ship is in port. Also, time underway when not deployed can be better utilized, so that there are fewer days away from home.

The Navy could consider increasing compensation. As noted above, many sailors would accept an additional year or two, or even another sea tour, for additional pay. The most effective tool is almost certainly sea pay, which could be targeted to those at the end of their sea tours. Not all need to accept this offer for the Navy to cut into the number of required shore billets.

Another option to consider is including the employment of military personnel as part of private contracts. That is, the contractor would agree to use some of the personnel in the performance of the service to the Navy. There are several benefits and limitations to this approach. First, if the Navy outsources to a high-quality provider, it will benefit from the training its personnel receive in leading business practices. Second, military personnel will eventually benefit when they leave the Navy because they can cite private-sector experience. To the extent that there are savings, they result from the Navy having fewer people in support activities. If a contractor is using military personnel, the Navy will still pay for them directly or, if the contractor pays part of the military personnel cost, the Navy will pay indirectly through the contract price. Also, this option does not address the drawdown in ashore requirements resulting from other efficiencies, such as regionalization, Smart Base, and direct vendor deliveries.

Another way to keep some of the military shore billets is for the Navy to include military personnel in their MEO and participate in the competitions. When military positions are now competed, they are automatically converted to civilian positions for the in-house bid. To keep some military positions in the in-house bid would require more flexibility in switching funds between Navy military personnel (MPN) and O&M. However, the evidence is that contractors will win most of these competitions, and only a few billets will be saved. Also, this approach alone does not help with other labor-saving initiatives.

Another solution would integrate military personnel with government civilian personnel. The Navy has started to do this in maintenance, where the civilian depot work force and the military intermediate maintenance work force are being combined. The merger allows for downsizing through the reduction in excess capacity and scale economies. It also may improve the productivity of both the civilian workers, who bring to the job greater maintenance expertise, and the military workers, who bring to the job the understanding of operational usage and diagnostics. More importantly, this may allow the Navy to cut civilian workers in lieu of its military workers. This approach works because of the excess capacity at the depots. However, just as there is a hidden cost in using military personnel instead of private civilians in many areas, it is also likely that there is a hidden cost in using military personnel instead of government civilians in many jobs.

In the long run, the Navy needs to cut personnel at sea. Programs such as Smart Ship can cut into those requirements. Cutting those billets not only reduces shore billets reserved for sailors, but it also cuts into the total number of support personnel needed.

Infrastructure efficiencies can reduce the need for facilities and the associated construction costs. They can also cut inventories. But the biggest costs are in personnel, and it is those costs that must be cut if the Navy is to shift resources from support into modernization.

A-76 Competition Practices

In its visit to the San Diego naval bases, the committee was briefed on successful public-private competitions (e.g., family service centers) and a private-private competition for third-party deliveries of office supplies throughout the area. Drawing on these experiences, and those of others,[20] the committee offers its suggestions on ways to improve the numbers and quality of local A-76 competitions.

First, the Navy should remove the local obstacles to effective competition. Today, managers of local activities have little or no reason to downsize or compete unless directed. Currently, installation managers must (1) pay for conducting the analyses leading to the competition, (2) incur the inconvenience or hardship of losing employees and dislocation, and (3) return the savings realized to the U.S. Treasury—all while being measured on the level of service provided.

Second, the Navy should focus on larger competitions, anticipating the results of regionalization, by carrying out process redefinition and holding competitions concurrently. For example, if a larger number of qualified bidders could be solicited, functions might be combined, since experience shows that the percentage of savings per billet increases with the size of the function competed.

Third, those in the chain of command must unequivocally support the competitive process. Because the supporting analyses of solicitations take several years in most cases, formalized organizations charged with planning and implementing major competitions will be required to provide continuity as the leadership rotates.

---

[20]Trunkey, R. Derek, Benjamin P. Scafidi, Francis P. Clarke, Cheryl Kandaras, Andrew M. Seamans, LCDR Carolyn M. Kresek, USN, Robert P. Trost, Angela L. King, Christine H. Baxter, Kerensa E. Riordan, Steven Smith, and Michael Ye. 1998. *Moving Forward with A-76 in the Navy,* CRM98-9.10, Center for Naval Analyses, Alexandria, Va., April.

Findings

In the areas of competition and sea-shore rotation, the committee's key findings are the following:

• The Navy's decentralized tactical process to identify A-76 competition candidates and to execute the program is not producing the large competitions necessary to reach the Navy's goal. Larger, more strategically executed competitions are needed to attain the Navy's competitive sourcing goal.

• Larger competitions have traditionally produced greater initial savings. In the past, competitions for fewer than 10 positions produced estimated initial savings of 22 percent; competitions for activities with more than 200 positions produced savings of 45 percent.

• *Supporting findings:* There are still significant opportunities to reduce costs with competitive sourcing.

— According to the Navy's records, there are 300,000 commercial activity positions (of which 60 percent are civilian) and, at most, only 10 percent have been previously competed. Even if all traditionally noncompeted work (such as depot maintenance and R&D) were excluded, the Navy would still have 55,000 civilian positions and 17,000 military positions available for competition.

— Using the estimated savings from the 1980s competitions and adjusting them for changes in the mix and size of activities and the civilian-military makeup of the work force would suggest slightly lower initial savings: from 31 percent to 27 percent. However, to the extent that there are trends, estimated savings have increased over time. If competitions are arranged by the year in which they were decided, the estimated percentage saved is higher in the later years. Also, the drawdown in forces and manpower seems to have had no effect on the percentage saved. The DOD work force peaked in 1987. The estimated percentage saved continued to increase into the early 1990s, when the earlier program ended. The most recent competitions are producing estimated savings greater than the historical 30 percent average. In the Navy, each of the three Navy competitions completed in 1997 and 1998 had estimated savings of more than 35 percent. The Air Force reports that it is now initially saving an average of 34 percent. Other DOD efficiency initiatives are either increasing or not affecting the savings from competitions. For example, the Navy regionalized and competed three family service centers concurrently in the San Diego region, an approach that produced greater savings than would have resulted from three separate competitions. Also, the Navy has used efficiency reviews in an attempt to reduce costs. Although the internal efficiency review at the telecommunications facility in Stockton, California, produced no savings, a later competition produced an estimated initial savings of 37 percent.

— Many military personnel remain in commercial activities, and personnel policies are a major limit on the number of billets available for competition, with only 16,800 of 138,000 military shore billets available for competitions under the implementation of current policies on sea-shore rotation and homebasing. The Navy has limited its own flexibility to target activities and consolidate them into larger competitions.

Conclusions

Past uses of competition, for the most part, are based on the practices of the 1980s, which were far from optimal. Recently, there has been a revolution in strategic sourcing in the commercial world. Moreover, although some shore jobs must remain as military positions, the Navy needs to better reconcile its infrastructure efficiency and seagoing personnel management objectives. This should increase opportunities for competition. It is the committee's conclusion that the Navy could increase savings further through actions such as the following:

- *Taking a more strategic approach to competitions.* The Navy can bundle positions for competition, select sources as long-term partners, and induce continuous improvement to match the practices of innovative commercial firms.
- *Regionally competing support functions.* The Navy now can take advantage of scale and scope economies. Half of the competitions between 1978 and 1994 were for 14 or fewer positions. Competitions of that size are relatively costly to administer, do not attract the most successful firms in the private sector, do not provide opportunities for economies of scale or scope, and produce the least savings. In the past, 37 percent of competitions for 10 or fewer positions produced no savings, whereas only 4 percent of competitions for 100 or more positions produced no savings.
- *Requiring OPNAV approval to cancel competitions.* Many competitions are being canceled by command-level authority.
- *Continuing to critically review inherently governmental positions.* Many of these jobs could be commercial, and the use of this exemption category varies considerably across locations and parts of DOD.
- *Removing disincentives to competitive sourcing.* Until recently, local commands paid for the competitions but did not keep any of the savings. Headquarters should pay the cost of the competition and allow local commands to keep some of the savings to improve facilities and quality-of-life services.
- *Adjusting military personnel policies to allow for competitive sourcing of more continental United States (CONUS) shore positions.* For example, the Navy includes sea billets normally filled by first-termers to justify shore billets for its careerists. In some cases, additional pay may be required to attract personnel to longer operational tours or shorter CONUS tours. Given that planned savings from competing functions performed by military personnel far exceed those from competing functions performed by civilians, targeting additional pay

to those affected could be a cost-effective approach to increasing the number of military positions available for competition.

- *Creating competition support offices as corporate repositories of lessons learned and providers of expertise.* Regional and local commands will be less frustrated if they know how others have solved problems similar to their own.
- *Committing strong leadership to overcome resistance throughout the Navy and DOD.* Large-scale use of competition requires a commitment from those at the top.
- *Making provisions for sea-shore rotation slots in any competitive sourcing.* That is, Navy personnel could work in contractor-managed billets for their shore rotations. To increase retention, contractors could be prohibited from hiring military personnel until they had completed another sea rotation.

Major Recommendation

The Navy should use competitive sourcing as a preferred approach to selecting the best providers of all support. In this regard, the Navy should establish a cross-functional team under the Assistant Secretary of the Navy/Installation and Environment (ASN/I&E) and the Assistant Secretary of the Navy/Research, Development, and Acquisition (ASN/RDA) to be responsible for overseeing the execution of competitive sourcing in business operation areas approved by Navy leadership.

In addition, the Navy should address all existing constraints in sea-shore rotation. The CNO should broaden the objective to managing seagoing personnel as a part of total naval personnel management and should direct relevant elements of the Office of Naval Operations (OPNAV) and second-echelon commands to search for innovative ways to satisfy the morale and retention needs that allow greater flexibility in reducing the cost of the infrastructure.

Recommendations—Competition

- *The SECNAV-OPNAV-fleet team should determine the best approach for each business operation area with respect to stopping provision of services in-house and instead establishing private-private competitions, public-private competitions, government-owned contractor-operated facilities, or public-private partnerships.*
- *The Navy should no longer depend on a bottom-up tactical approach. The senior cross-functional team should recommend to the CNO and SECNAV particular areas in which to stop in-house provision and instead consolidate into large, more strategic competitions.*
- *The CNO should develop a schedule for the actions recommended above that reflects the urgency of reducing costs in business operations.*

Recommendations—Managing Seagoing Personnel

- *The CNO should expand the Navy's current objective beyond sea-shore rotation to finding the best ways to attract and retain sailors.*
- *The CNO should task an appropriate independent analysis and research firm to better define the relationships among sea tour lengths vis-à-vis shore tour lengths, sea pay, time with family, work satisfaction, and other appropriate factors, to determine the impact on morale and retention of seagoing personnel.*
- *The CNO should establish an ongoing dialogue between the CNO staff (including all stakeholders in reducing the cost of the infrastructure) and the fleet commanders to better define the specific attributes of maintaining morale and retention of seagoing personnel in the Navy.*
- *The CNO should direct relevant elements of the OPNAV and second-echelon commands to (1) search for new and innovative ways of satisfying the morale and retention requirement that would allow greater flexibility in reducing the cost of the infrastructure, and (2) carefully review the requirement or policy for the shore duty billets to be closely associated with the skill(s) required at sea. In addition, the committee recommends the following: (1) balance the potential gain or loss of skills with the loss in flexibility in accommodating new opportunities for privatization of functions; (2) strive to obtain better integration of shore duty billet policies with larger Navy programs and competitive sourcing initiatives; (3) continue and expand the merging of intermediate- and depot-level maintenance programs with military personnel integrated with civilians in shipyards and NADEPs; and (4) look for ways to make use of military personnel in other contractor facilities and in competitive sourcing contracts.*
- *The CNO should change the veto authority of the N1 over billets nominated for competitive sourcing to one of greater management consensus by all of the stakeholders, with the VCNO as the final arbitrator.*

IMPLEMENTING THE STRATEGY

Changes of the kind and magnitude visualized under the strategy outlined in this chapter do not come easily to a large, tradition-based organization. In part, the changes will be induced by the new Navy systems that are acquired, in which new technologies and new system designs will make their own demands on the support system. However, the Navy cannot wait to implement infrastructure change with new acquisition programs. An intensified campaign to make the most of the opportunities presented, especially when it entails extensive organizational and budgetary shifts, will be challenging in the extreme and should be initiated now. The changes involved will entail risks to achieve the benefits sought, and managing—not minimizing—risks will be essential to success.

The benefits to be achieved are obvious. The portion of the Navy's personnel, material, and other resources available to apply to the Navy's core systems and missions will be increased. The Navy's use of highly trained personnel can be better applied to those systems and missions. In improving shore support, the quality of life of Navy personnel and their families will be improved, leading to higher rates of retention and better-trained crews. Overall fleet readiness and combat capability will be improved within given, constrained budget levels. Also, tendencies toward isolation of the Navy from the civilian economy and population that may have attended the shift to an all-volunteer, career force and reduction of its supporting infrastructure through the BRAC process will be lessened or reversed.

The risks are those attending disruption and establishment of new organizational connections and procedures. Traditional organizational responsibilities and authorities will be shifted to others who will have to learn how to use them in new circumstances. Unfamiliar funding patterns will require different justification that may be more difficult to support with budgeting authorities, such as OMB and the Congress, to which the new patterns will be unfamiliar. The order of sequentially dependent changes may be lost in the processes of overcoming institutional resistance, throwing the effectiveness of such changes in doubt without remedying the negative effects of the disruptions or gaining all of the positive benefits that were sought. Mistakes will be made. Some may be costly, using resources sorely needed for substantive investments; others may be irretrievable in a less forgiving oversight environment, causing ripples that could further disrupt the process of change.

In the last analysis, all such risks entail potential added expenditures of time and resources in effecting the sought-after changes, with the chance of using the fleet's financial resources ineffectively and, in consequence, hurting its readiness and combat capability. The benefits, to the extent that they are achieved, will greatly enhance resources available for the Navy's core missions and capabilities, and its readiness along with them.

It is the balance between the risks and rewards that must be weighed by Department of the Navy decision makers in taking each step of the strategy to rationalize the fleet system and its support component. Perhaps the only certainty in the equation is that failure to take at least some of the risks will lead inevitably to further decline in the fleet's efficient and effective use of any available resources, and consequently in its size, capability, and readiness. There thus seems to be no alternative to going forward, managing the risks as they are anticipated and felt.

4

The Need for Top Leadership to Drive Change Across the Entire Navy System

THE COMPELLING CASE FOR MAJOR CHANGE ACROSS THE NAVY

Current Navy infrastructure business initiatives (i.e., regionalization, Smart Base, competitive sourcing, and so on) are good starts, but they have produced only a small fraction of the funds needed to meet the Navy's modernization goals. The current efforts, achieved largely by allocated budget reductions, have taken about $0.75 billion per year out of the infrastructure, but a reduction in the range of $3.5 billion to $5.0 billion per year is needed to return the modernization budgets to the level deemed appropriate by the Navy.

The budget reductions achieved to date appear to be arbitrary and isolated, the result of tactical actions. The committee could not identify an overall corporate Navy strategy to integrate, prioritize, and allocate resources so as to achieve essential fleet support at significantly less cost for the support elements of the total naval system. Also, the compelling case for major change in the way the Navy conducts its infrastructure business is not being made by the senior leadership. Change is being led by isolated support and staff elements who lack the directive and authority to impact the total Navy system. No proven methodology for conducting the evaluation of the current system and for implementing and managing the corporate Navy change could be identified by the committee.

Thus, much more dramatic change, including a major cultural change that only the Chief of Naval Operations can lead, will be needed if the Navy is to succeed in deriving sufficient modernization funds out of its infrastructure. Failure to act quickly and comprehensively to bring infrastructure costs down will likely result in an inability to acquire some of the modern ships, submarines, aircraft, and information systems that will keep the U.S. Navy both at its current size and preeminent in quality in the future.

The magnitude of the change required is very large and will be difficult to achieve. The difficulty is compounded by the fact that the Navy is a tradition-driven organization that has changed only slowly over the years. The entire Navy organization will have to reengineer itself to become more businesslike and cost-efficient. This will require senior leadership to engage *all* elements of the Office of the Chief of Naval Operations (OPNAV) and second-echelon commands in the change process. Since the fleet forces levy the requirements on the shore-based and support organizations, all of these groups must engage in a consensus-building change process that will establish the desired balance between forces, modernization, and infrastructure. The comprehensive change process must begin immediately.

Significant change, unfortunately, is not a natural process for most people. Most people become very comfortable with the present state that they have helped create. Their contributions have established their sense of self-image and worth to the organization and of their organization to the world. Even when situations and conditions arise that demand rapid, responsive change, such as the budget crises within the Navy, many people are reluctant to take action because they (1) refuse to recognize the seriousness of the threat, (2) are not sure what to do, (3) do not want to give up what they currently possess, or (4) fear the unknown future state. Change must be managed by senior leaders who have a compelling reason to move to a new state, have a clear vision of that future state, and have a plan to deal with the obstacles and impediments that stand in the way of that future.

Fortunately, there are numerous examples of organizations that have successfully managed major change in response to threats to their existence. There are key elements for success that have been derived from these cases. The processes or methodologies used by successful leaders to bring about permanent, lasting change within their organizations have also been identified.

Unfortunately, numerous examples of failure to change quickly and decisively also exist. U.S. automakers in the 1980s failed to recognize the serious threat from Japanese automakers and lost significant worldwide market share that has not been recovered. U.S. commercial shipbuilders failed to deal with the serious cost threats from Japanese, Korean, and other shipbuilders and have essentially become minor players in the world market. IBM failed to recognize and deal with the threat that minicomputers brought to the mainframe computer business and required many years to rebound. Digital Equipment Corporation failed to recognize quickly enough the impact that personal computers would have on the minicomputer business. As a result, Digital Equipment Corporation has been consumed by Compaq, a personal computer manufacturer. The consequences of failure to act quickly can be significant.

In this chapter, the committee highlights the vital role of leadership in making change happen. A set of key guides or elements to any successful change is given in Appendix D. An example of a proven methodology for taking large organizations through major change is also provided. The senior leadership of the Navy should use one of these proven step-by-step processes to rapidly move the organization to the desired future state.

THE LEADER'S CRUCIAL ROLE IN CHANGING THE ORGANIZATION

The leader of any major organization plays a pivotal role in creating the successful future state. For the U.S. Navy this leader must be the Chief of Naval Operations (CNO) with the support of the Secretary of the Navy (SECNAV). This role cannot be delegated downward if the entire organization is to change. Only the CNO has the authority to bring the fleet elements and the support naval units to the change process and to demand that results happen. Only the SECNAV has the authority to bring all of the stakeholders in the Department of the Navy to the change process and to demand that results happen. Change will occur below this level if led by the leader of a subordinate organization, but it will be confined largely to that organization and fragmented for the Navy as a whole. In studies of organizations that have made significant and lasting change, the leader of the organization has been observed to display some specific characteristics described in the sections below.

Be the Champion for Change

The leader must be the champion for change within the organization. This role cannot be relegated to direct reports or to staff personnel. The reasons and urgency for change must be communicated clearly and often. The reasons must be compelling to the people in the organization, and the negative impacts of not making the required changes must be put forth. The leader must demand that direct reports follow his or her example in driving the change down through the organization. All personnel within the organization must clearly recognize the importance and high priority that the leader places on the need to change. The leader may designate other champions within the organization to reinforce, facilitate, and implement change, but there is no substitute for the top leader being perceived as the driving force behind the change process.

Create the Vision of the Future State

The leader must develop the vision of the future state of the organization. For the operational Navy, this must be the CNO in conjunction with senior leaders. This must be a clear and consistent vision that is communicated to *all* within the organization. It must be a future state (e.g., 5 years) that is more desirable than the present and one that members of the organization deem desirable. The CNO must convince others that the future state is achievable through hard work and smart decisions. For example, this future state could be a smaller but more capable and ready Navy in which global connectivity provides information superiority and is coupled with smart weapons to provide dominance in naval warfare.

In essence, when a clear vision of the future state 5 years down the road is communicated, it becomes the "stake in the ground" from which planning backwards to the present state can be accomplished. Although many in the Navy have a vision of the future, only the CNO has the authority and resources to make the change to a future state happen for the entire Navy organization. With a clearly stated and communicated vision from the top, other visions within the organization can be aligned and made consistent with the overall vision of the Navy. Efforts to create and align visions and change at lower levels in the absence of this unifying, overarching vision and action will at best be fragmented and frustrating for the individuals involved.

Set Strategic Goals

The CNO with senior leaders, and with the concurrence of the SECNAV, must set the strategic goals of the operational Navy. The goals must be of the highest importance to the future direction of the Navy. For example, a strategic goal might be that the Navy will be capable of engaging in and winning two wars simultaneously anywhere on the globe. Strategic goals must be clear and measurable. They should be "stretch" goals, that is, ones that can be accomplished only through sustained hard work. They might be expressed in terms of numbers of ships, submarines, and aircraft in existence in 5 years in order to conduct two wars simultaneously. They might be expressed in terms of the budget available for specific modernization through reductions in other areas. For example, a strategic goal might be that over the next 2 years the Navy will reduce total infrastructure costs by 20 percent through process changes and competition, with those savings being allocated to the funding of DD-21, IT-21 initiatives, and the joint strike fighter (JSF).

These strategic goals will set in place the tactical plans and actions that will lead to accomplishment of the desired results. When the strategic goals are accomplished, the organizational vision will be realized. When goals are vague, such as "achieve cost reductions without sacrificing performance," success cannot be measured and the results will not equate to the vision.

Involve All the Stakeholders

The leader is the only person who has the authority to involve all the stakeholders in the change process. In any organization involving multiple, dependent functions, the leader must bring together and engage all of the participants. For the operational Navy, the CNO must be the person to engage all of the stakeholders in the change process. This involves the fleet commanders, the infrastructure commanders, the systems commanders, and the CNO staff.

At the next level of the Navy, fleet operations involving the surface, submarine, and aircraft elements, and the shore-based elements, including the systems commands, form a closed system. The warfighting elements place requirements on technology, weapons, logistics, shore housing, job rotation, and so on provided by the support elements. These requirements must always be prioritized and traded to realize the strategic goals. With reduced budgets, these trade-offs are even more critical. The fleet commanders must engage the heads of *all* of these elements in the change process. It is natural that not all participants will want to engage in change equally. It is the role of the fleet commander in this case to demand participation from all parties critical to achieving the future state of the organization.

Establish the Framework and Process for Involving All the Stakeholders

The leader at any level must establish the framework that will be used to manage the change process. Meetings should be scheduled at regular intervals. Attendance by all of the stakeholders must be mandatory. The leader must be in attendance and guide the discussions. Unless a high priority is placed on these meetings and the leader is in attendance, the chances of achieving difficult change in which compromises and concessions are necessary are slim.

Employing a proven end-to-end process for managing change will greatly facilitate these meetings. Fortunately, there are many proven processes available. The elements of one such successful process are described in Guide 7 in Appendix D.

Allow Mistakes

As the members of the organization undertake change, mistakes will be made. In an organization such as the U.S. Navy in which following procedures has been sacred, departures from the norm and mistakes have been viewed as unacceptable. As a result, members of the organization are reluctant to take risks. In order to encourage innovation and change, the leader must make it clear that mistakes are an opportunity to learn and to start again along the path to permanent change.

Demand Results on a Timetable

The leader must demand results on a timetable commensurate with achieving the goals and vision. Change is difficult, and it is unnatural for humans to give up authority, control, budgets, and so on for the greater good of the organization. Unless the leader demands that specific goals be achieved on a timetable and provides the tools to make the changes, procrastination and only minor accomplishments will be realized. Visible rewards for those who meet and exceed goals will reinforce the change.

Communicate

The leader must constantly communicate the vision and strategic goals to all members of the organization and present a compelling case for why everyone in the organization needs to buy in and become part of the change process. Forums, broadcasts, newspapers, and luncheons are all mechanisms for spreading the word and developing understanding, cooperation, and participation. Changing a culture is difficult and takes a long time to accomplish. The leader must be the great communicator in convincing people to join in the path to a brighter future.

Finding

The committee recognizes that many cost reduction initiatives are underway, but the compelling case for major change in the way the Navy conducts its business is not being made by the senior leadership. Change is being led by isolated support and staff elements who lack the directive and authority to impact the total Navy system.

Major Recommendation

To achieve its recapitalization funding goal, the Navy should develop and implement a corporate-wide strategy to improve the business operations of the entire Navy infrastructure. The senior leadership of the Navy, led by the Secretary of the Navy and the Chief of Naval Operations, should establish a clear vision and a corporate-wide strategy for conducting the future operations of the entire naval system within the budget constraints projected. The strategy must be clear on what is to be achieved, in concrete terms, how it is to be achieved, with what means it is to be achieved, and when it is to be achieved. The strategy, of necessity, must address all portions of the Navy infrastructure, not just a few isolated portions thereof. A compelling case for major change in the way business is conducted must be made by the CNO and communicated to all elements of the Navy. Responsibilities and authorities to implement change must be made clear and issued by the CNO.

Recommendation

The CNO should make participation in the change process mandatory for all elements of the Navy. Target reductions, implementation calendars, and progress monitoring should be established. Results should be demanded and organizations and individuals made accountable. Innovative ways of conducting business should be encouraged, supported as appropriate, and rewarded based on measurable results.

Appendixes

A

Admiral Johnson's Letter of Request

CHIEF OF NAVAL OPERATIONS

NRC/CPSMA

25 April 1997

Dear Dr. Alberts,

In December 1995, at the request of this office, the Academy's Naval Studies Board initiated a study entitled "Technology for Future Naval Forces." As this effort draws to a close, I look forward to receiving the results of the study.

In addition to your current work, I would like you to consider undertaking two additional studies in the areas of "Improving Shore Installation Readiness and Management" and "Naval and Expeditionary Logistics Innovation." My staff will develop terms of reference for these two one-year studies in consultation with the Chairman and Director of the Naval Studies Board.

Thank you for all your support. I value our continuing and close working relationship with the National Academy of Sciences.

Sincerely,

JAY L. JOHNSON
Admiral, U.S. Navy

Dr. Bruce M. Alberts
President
National Academy of Sciences
2101 Constitution Avenue, N.W.
Washington, DC 20418

B

Analysis of Department of the Navy Funding Trends

The overall Department of the Navy funding by Service is depicted in Figure B.1.[1] It shows a significant decline in funding for both the Navy and the Marine Corps from FY 1991 through FY 1994 and then a leveling out through FY 1999.

The Marine Corps share of the Department of the Navy budget has increased from 13.5 percent in FY 1991 to 16.2 percent in FY 1999 and continues to increase at an average rate of 0.3 percent per year, as shown in Figure B.2.

The funding for the overall Department of the Navy and for the two Services is depicted in Figures B.3, B.4, and B.5.

Department of the Navy funding was broken down into three categories: (1) infrastructure using the OSD (Program Analysis and Evaluation [PA&E]) definitions of infrastructure; (2) force operations using the O&M and military personnel appropriations in the OSD (PA&E) definitions of forces; and (3) force modernization using the research, development, test, and evaluation (RDT&E), procurement, and military construction appropriations in the OSD (PA&E) definitions of forces.

In almost every case there is a clear difference in funding trends between FY 1991-1994 and FY 1995-1999. The funding trends appear to show that force modernization bore the brunt of the overall budget reductions and then recovered

[1]Unless otherwise noted, all data shown in this appendix are a result of the committee's use of FY 1999-2003: President's Budget Future Years Defense Programs. In addition, please note that all data are shown in constant FY 1999 budget dollars: FY 1991 through FY 1997 are actual obligations, FY 1998 is the current budget in execution, and FY 1999 is the President's budget currently being considered by Congress.

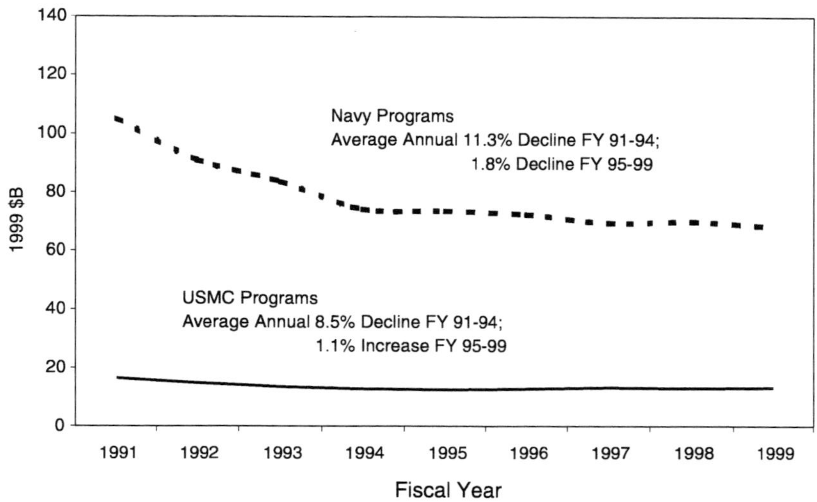

FY	1991	1992	1993	1994	1995	1996	1997	1998	1999
Navy ·······	105.0	91.0	83.6	74.2	73.5	72.4	69.5	70.0	68.3
USMC ———	16.4	14.8	13.5	12.8	12.5	12.8	13.2	13.0	13.2
Total	121.4	105.8	97.1	87.0	86.0	85.2	82.7	83.0	81.5

FIGURE B.1 Funding for Department of the Navy programs, 1991-1999.

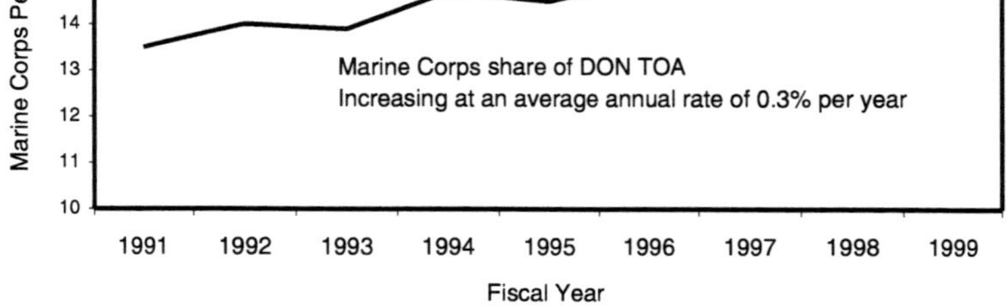

Fiscal Year	1991	1992	1993	1994	1995	1996	1997	1998	1999
USMC ———	13.5	14.0	13.9	14.7	14.5	15.0	16.0	15.6	16.2

FIGURE B.2 Marine Corps percentage of Department of the Navy (DON) total obligational authority (TOA), 1991-1999.

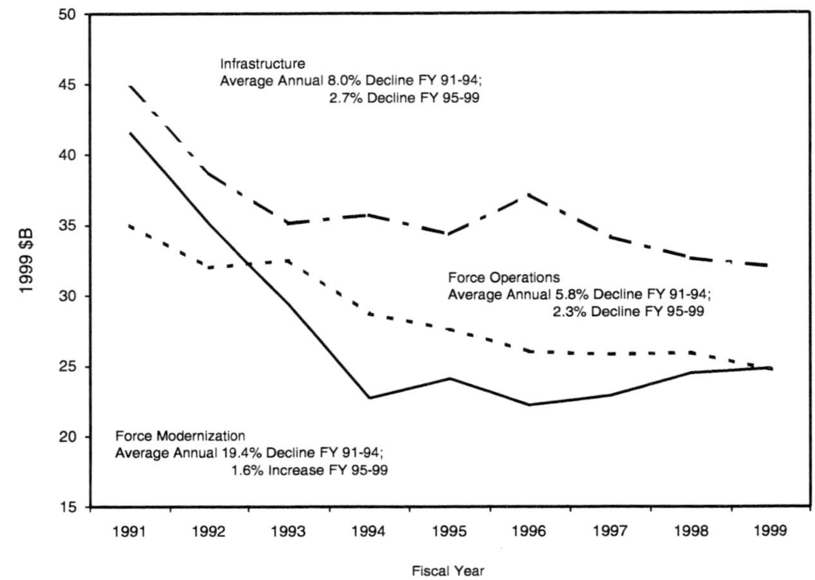

Fiscal Year	1991	1992	1993	1994	1995	1996	1997	1998	1999
Modernization ▬▬	41.6	35.1	29.4	22.7	24.1	22.2	22.9	24.5	24.8
Operations ·······	35.0	32.0	32.5	28.7	27.6	26.0	25.8	25.9	24.7
Infrastructure ▬ · ·	44.8	38.7	35.1	35.7	34.3	37.1	34.1	32.6	32.0
Total	121.4	105.8	97.1	87.0	86.0	85.2	82.8	82.9	81.5

FIGURE B.3 Funding for Department of the Navy programs, 1991-1999.

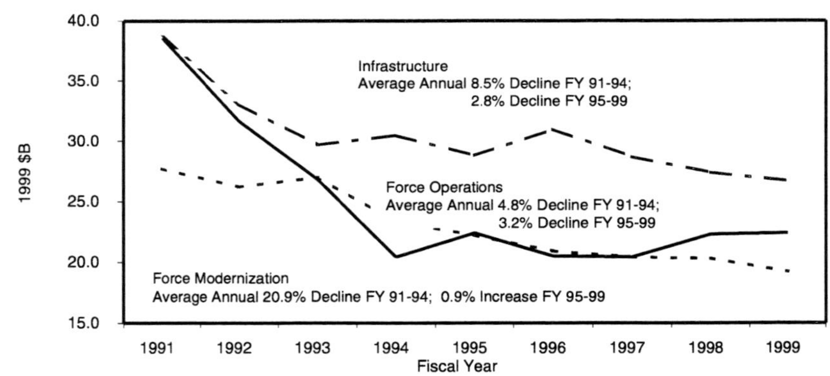

Fiscal Year	1991	1992	1993	1994	1995	1996	1997	1998	1999
Modernization ▬▬	38.6	31.7	26.8	20.4	22.4	20.5	20.4	22.3	22.4
Operations ·······	27.7	26.2	27.0	23.3	22.2	20.9	20.4	20.3	19.2
Infrastructure ▬ · ·	38.7	33.1	29.7	30.5	28.8	31.0	28.7	27.4	26.7
Total	105.0	91.0	83.5	74.2	73.4	72.4	69.5	70.0	68.3

FIGURE B.4 Funding for U.S. Navy programs, 1991-1999.

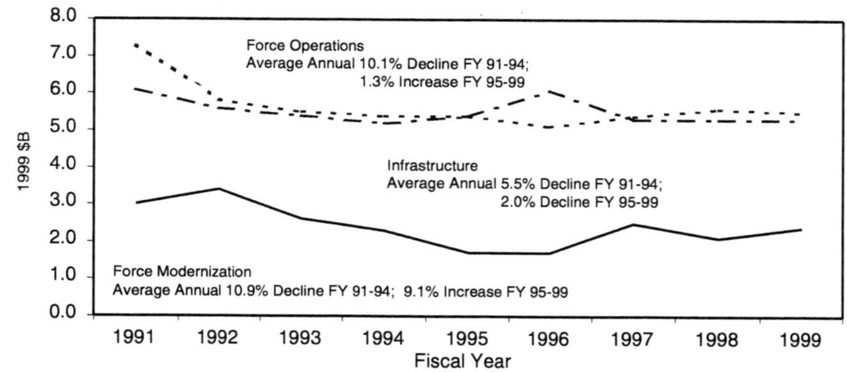

Fiscal Year	1991	1992	1993	1994	1995	1996	1997	1998	1999
Modernization	3.0	3.4	2.6	2.3	1.7	1.7	2.5	2.1	2.4
Operations	7.3	5.8	5.5	5.4	5.4	5.1	5.4	5.6	5.5
Infrastructure	6.1	5.6	5.4	5.2	5.4	6.1	5.3	5.3	5.3
Total	16.4	14.8	13.5	12.8	12.5	12.8	13.2	13.0	13.2

FIGURE B.5 Funding for U.S. Marine Corps programs, 1991-1999.

slightly from FY 1995 to 1999. Navy programs follow a similar pattern, but the magnitude of the reductions is somewhat greater and is depicted in Figure B.4. Marine Corps (USMC) programs have smaller declines and larger increases and are depicted in Figure B.5.

Table B.1 depicts the results of an analysis to determine if and to what extent there were resource transfers between programs and Services during the period between FY 1991 and FY 1999. The first column lists the categories; the second column is FY 1991 funding in billions of FY 1999 dollars; the third column is FY 1999 funding in billions of dollars; the fourth column shows the reduction in funding from FY 1991 to FY 1999; the fifth column lists the percent reduction by category from FY 1991 to FY 1999; the sixth column indicates the FY 1999 funding that would have occurred if each category were reduced by the same Navy Department overall percentage; and the seventh and last column represents the difference between actual FY 1999 funding and the normalized FY 1999 funding. This last column indicates the degree to which a category had funding transferred either into it (a positive number) or out of it (a negative value).

This analysis shows that Navy modernization was the bill payer for the overall Navy Department budget reductions and resulted in an excess $3.5 billion reduction that must be recouped from other programs in order to reestablish the former funding relationships of about a decade ago and continue capitalization of the Navy. If Navy modernization requirements exceed the former share

TABLE B.1 Resource Transfers Between Programs and Services (in billions of FY 1999 dollars)

Category	FY 1991	FY 1999	FY 1991 to FY 1999 Reduction in Funding	FY 1991 to FY 1999 Percent Change	Normalized FY 1999	Transfer
Department of the Navy						
Modernization	$41.6	$24.8	$16.8	40	$27.9	–$3.1
Operations	$35.0	$24.7	$10.4	30	$23.5	$1.2
Infrastructure	$44.8	$32.0	$12.9	29	$30.0	$1.9
Total	$121.4	$81.5	$40.0	33	$81.5	$0.0
Navy						
Modernization	$38.6	$22.4	$16.2	42	$25.9	–$3.5
Operations	$27.7	$19.2	$8.5	31	$18.6	$0.6
Infrastructure	$38.7	$26.7	$12.0	31	$25.9	$0.7
Total	$105.0	$68.3	$36.7	35	$70.4	–$2.1
USMC						
Modernization	$3.0	$2.4	$0.6	20	$2.0	$0.4
Operations	$7.3	$5.5	$1.8	25	$4.9	$0.6
Infrastructure	$6.1	$5.3	$0.8	14	$4.1	$1.2
Total	$16.4	$13.2	$3.3	20	$11.0	$2.1

NOTE: Totals may not add due to rounding.

of the Navy Department budget or it is necessary to continue to increase the Marine Corps share of the budget, the $3.5 billion figure must be adjusted upward accordingly.

For the reasons articulated in Chapter 1, the committee focused its review on the approximate half of the Navy infrastructure that is more closely associated with the N4 organization (i.e., those portions of the infrastructure concentrated on base closure and environmental compliance, installations (less base closure), central logistics, and quality of life). The total Navy infrastructure and the portion initially considered by the committee are depicted in Figure B.6.

The committee also looked at infrastructure by appropriation since different offices in OSD and different committees in Congress review different parts of the budget by appropriation. Not surprisingly, the O&M account and the MPN account make up the majority of the funding for infrastructure. This result is displayed in Figure B.7 for both the total infrastructure and the portion reviewed by the committee.

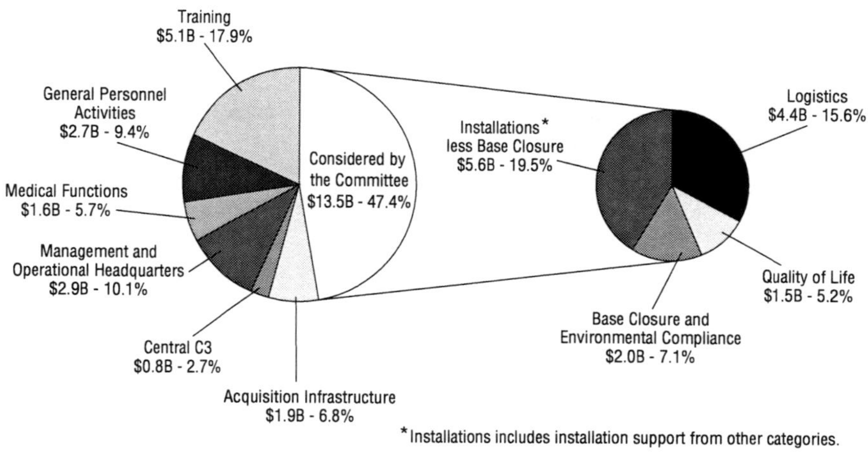

FIGURE B.6 Navy infrastructure by functional categories initially considered by the committee. Data shown are average values from FY 1995 to FY 1999.

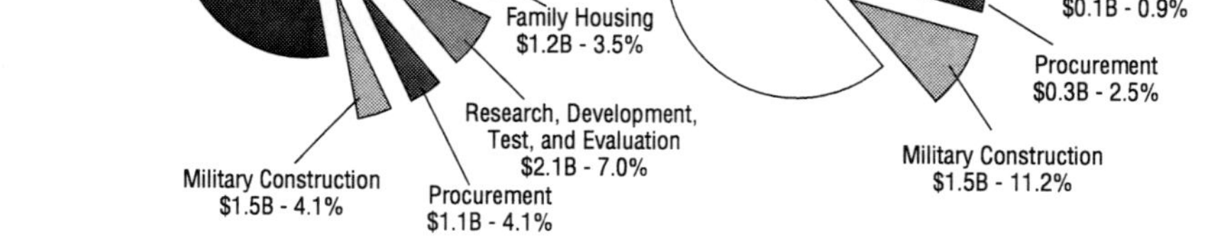

FIGURE B.7 Navy infrastructure by appropriation initially considered by the committee. Data shown are average values from FY 1995 to FY 1999.

C

Developing Output Measures

Step 1—Segment the Infrastructure into Discrete Functions

Starting within each region and with each program manager, define discrete nonoverlapping and mutually exclusive functions at each facility or base that has a single person who is both responsible and accountable. This function (e.g., security) should have definable boundaries and be easily recognized. The sum of all functions should comprise the universe of all goods and services to be provided by the region.

Step 2—Identify the Users of the Function

The users not only should be customers but also should have the ability to "vote" on the performance of the function and agree to the ranges and limits of the function to be performed. For example, in port operations the individual ship's commanding officers are the proximate users of the function, but only the type commanders can agree to the range and limits of port operations services to be provided and potential restriction on ships operations.

Step 3—Develop a Performance Work Statement for the Function

Although there may be no intention of privatizing this function, assume that some other entity will perform it and that one has to describe in detail what must be performed, including attributes of timeliness, quality, and scope. This initial description should be made by the person who is currently responsible and accountable for the function and should be assisted by those performing the function. The performance work statement (PWS) of objectives should be formulated at three levels:

1. *Threshold.* This is the minimum level of performance and may require some restrictions and/or assistance from the user. For example, air operations at an outlying field may be restricted to certain hours of operations or a contingent from ship's force may be required to assist in security patrols. The specific restrictions and/or assistance should be spelled out in detail.
2. *Objective.* This is the expected or normal level of performance. It should be specific in terms of scope of services, timeliness, and so on, that define the goods or service provided. For example, berthing will be provided for up to six Aegis cruisers without nesting; 4,000 amperes of electrical power will be provided within 15 minutes of berthing; 25,000 gallons of fresh water will be provided per day at 30 pounds per square inch constant pressure; and so on.
3. *Enhanced.* This is a level of performance over and above the objective that may be required for special situations. For example, special security during special weapons handling or transportation might be provided.

Step 4—Performance Work Statement Cost

With an agreed-to statement of work and performance, it is now possible to determine the cost of the goods or service. This statement is broken down into three substeps:

1. *Activity-based spending (ABS).* This is often looked at as activity-based costing, but it is only the first step. Here a survey is made of what the institution is *spending* on a function. This establishes the differences in spending between various activities doing similar work and determines what the activity is spending resources on and what it is not spending money on. This is the first check on the performance work statement to establish that it is complete (i.e., one is spending resources on all of the elements within the PWS) and has not overlooked a necessary performance within the PWS (i.e., spending resources on something that is not in the PWS rather than something that one should be spending on). This leads to a possible revision of the PWS.
2. *Capitalization.* After it has been determined what an activity is currently spending on a function, an analysis should be conducted to determine if an additional investment in capital (e.g., a new information system, additional training) would result in a net lower cost of operations.
3. *Activity-based costing (ABC).* This takes ABS to the next step, which is determining what a function "should" cost. It can be accomplished in a number of ways:

- In situations where there is an industry standard or best practice, it can be interpolated to the current function.
- In situations where the same or similar function has been contracted to another claimancy or base, that contract can be interpolated to the current function.

Step 5—Performance Work Statement Review

Once the PWS "draft" has been completed, the users and major claimant should review it to determine if it meets their needs. In particular they should agree to or modify the PWS as follows:

- The threshold PWS is acceptable, and they are willing to accept the restrictions and/or assistance that is required if necessary.
- The objective PWS is what they expect in normal situations and what they are willing to support for funding.
- The enhanced PWS meets their requirements for special situations and what they are willing to pay for additional funding.

One of the principal benefits of this step is that the user and the major claimant "buy in" to the process.

Step 6—Prioritization and Reconciliation with the Budget

Together with the users, major claimant, regional commander, and program managers, all of the functions are prioritized by function and level of service (threshold, objective, enhanced) and compared with the budget or program across the Future Years Defense Plan (FYDP). Where there is a shortfall, either the priorities can be changed or, by going back to step 3, the PWS can be restructured. There is also the opportunity for the user to agree to shift resources from his or her budget to the region to procure required services.

Although this may seem to be a long and laborious process, it provides a clear definition of what goods and services will be provided, what they will cost, and what the real opportunities for savings are. There is also a clear definition of what capability is lost when the "requirement" exceeds the budget.

Step 7—Execution

Once a budget and level of service have been agreed to by all concerned, the person responsible for providing the service must be given the incentive to produce the service at the agreed-to cost. This may include personal performance reviews, opportunity for promotion and/or advancement, or sharing in any additional cost savings.

D

Key Guides to Successful Change

Beyond the leader's role there are some key guides to or elements of successful change. The guides synthesize lessons learned from previous (successful and unsuccessful) changes in the private sector and from many investigations of the way leaders intuitively go about instituting change. The guides summarize these intuitions and enable leaders to explain their reasoning processes. The guides also provide a framework for thinking about how to effect change and a set of checkpoints for any leader initiating a change process within a system. These guides will be useful to the senior leadership as they undertake the necessary process of change within the U.S. Navy.

The basic concept is that there must be overall change of an organization. Starting at the top is by far the best way to achieve major systemic change. Many people think that change "happens" from the top, when in fact it must be carefully "managed" from the top. Significant changes at any level should be considered desirable and should be sought even as overall organizational change is pursued. Overall organizational change very often starts in the trenches. That is, successes with change at a more local level, such as in the San Diego and Hampton Roads regions, are often motivations to accept larger changes in the system.

Everyone and every level of a system has to create its own innovations and changes in the way it works and conducts its activities, rather than waiting for

NOTE: For further reading on this subject, see Troxel, James P., ed. 1995. *Government Works: Profiles of People Making a Difference,* Miles River Press, Alexandria, Va., and Troxel, James P., ed. 1993. *Business Cases from Around the World,* Miles River Press, Alexandria, Va.

someone else to do it. The guides can be applied at each level of the organization, regardless of size. These guides form a basis for inculcating change as a common thread in all activities of the U.S. Navy, rather than responding only to the crises generated by reduced budgets.

Guide 1—Make a Compelling Case for Change

Significant reductions in the budgets available for modernization of the Navy are a very real basis for change. The reasons for the budget reductions, and the negative impact on the Navy if changes are not made to deal with these reductions, must be communicated to the whole organization. All personnel must understand the dire consequences of continuing business as usual. In the reduced budget environment, weapons systems will not be upgraded, platforms and bases will not be maintained properly, and modernization of the forces with information systems, such as IT-21, and new platforms, such as DD-21 and JSF, will not materialize.

The magnitude of the changes required to achieve the strategic objectives of the Navy within the expected budgets must be conveyed. For example, the fiscal reductions that are required in the infrastructure to fund the desired modernization goals of the Navy in a flat overall budget situation must be made clear. The CNO must identify a plausible, quantitative set of target reductions across the Navy that will achieve the strategic objectives. Once established, these targets must be communicated to the entire organization and a compelling case made that every element in the organization must do its part to make the required changes for the good of the Navy.

Opportunities for change should be identified continuously throughout the organization as a way of stimulating innovation and motivation for improvements. Successful systems and best practices in the Navy and from elsewhere can be communicated continuously to stimulate considerations of a "change climate." A continuing referral to the significant and continual change in the global business world is also a form of emphasizing the compelling need for continual change. The many and rapidly changing technological innovations that impact the Navy are another compelling case for embracing a continual need for creativity, innovation, and change.

The organization can be restructured to foster a compelling and continuing need for change. It can be made relatively flat to reduce bureaucracy. It can use cross-hierarchical and cross-functional teams to stimulate enterprise-wide behavior. The organization can reduce formalism to a minimum by reducing policies and procedures. It can foster communication and cooperation by providing a comprehensive and easily accessible information network. The organization should encourage more participative decision making and should use matrix assignments for people to introduce innovation.

Guide 2—Treat Each Situation Initially as Unique

One of the major barriers in achieving change at all levels of an organization is the belief that once a process, organization, technology, or new system is developed, others in similar situations will know that they should use it. This belief has caused huge difficulties in transferring change in all organizations because it fails to recognize the uniqueness of each location. The details of the Southwest Region solution, as good as they may be, should not be considered "the" solution for the Hampton Roads region or vice versa. Local people must in most ways create and buy into their own change process and solutions. The differences in people and culture do not mean that ideas and technologies from the first situation will not be used. Good ideas and technologies will be drawn, when identified as needed, from previous solutions and the best practices of others.

Guide 3—Put All Change in a Context of Larger Purposes and Missions

Even when a clearly stated purpose or mission is available, recognize that every purpose or mission must exist in a context of larger purposes or missions. Individuals make many minute decisions as time goes on in the change process, based on their *unstated* assumptions and the meanings associated with words and phrases. Focusing only on what may be considered a well-stated purpose leaves open for each individual the interpretation in his or her own way of the larger purposes that really ought to govern these minute decisions.

The operating concept for expanding the purposes is to reexamine the assumptions that are hidden in each person's mind. Discussing the larger purposes lets everyone understand in a nonthreatening way the different thinking of others so that group acceptance occurs. It assures all that the real question is being addressed rather than accepting the problem as stated. It focuses first on doing the right things rather than on doing things right.

Guide 4—Develop a Vision of the Future System to Guide Today's Actions

In effect, plan backwards from the future. The first part of implementing this guide is generating several alternative scenarios or options that satisfy the focus and larger purpose. Play with the scenarios, search for the operating dynamics, maximize the flow (remodel your mind, suspend judgment, use unusual media, seek the greatest value for the purposes, go back to zero, be absurd, and seek ways to eliminate the focus purposes by asking how to achieve the larger purposes).

From these major alternatives, it is possible to select a future state solution that should be sought, in say, 5 years. Focusing on this future state identifies the changes and revisions needed today to achieve the target. At this point, it is possible to determine if solutions are available elsewhere to save time and not "reinvent the wheel." The consideration of many options will provide some degree of assurance that changes needed in the future will not be blocked by what is done today. Having options available also provides a basis for contingency planning as external conditions change.

Being able to articulate the major alternative options from which the future solution target was selected is also very useful in making the recommendation for change to decision makers at all levels. The usual presentation to such decision makers puts them in the uncomfortable position of having to accept or reject the recommendation. Providing the options permits them to become part of the process of developing the solution and even to improve the recommendation and make it more workable.

Guide 5—Take a Systems Approach to the Change Process

Every recommended change, however big it may be, is always part of a larger system. As indicated previously, the CNO should lead the change process for the whole Navy, the largest system entity. Doing so will then allow the subsystem organizations to place their changes in the context of the larger system. Being able to show how the change will fit into and impact the larger system will produce a much greater probability that the change will be accepted. Any recommendation should include specific enough systems detail to show that it is workable (i.e., the users that are affected, the interfaces required, and so on).

Understanding this guide helps put changes in infrastructure into terms of specific system elements. The elements of the force operations can be related to force modernization and to the infrastructure to provide insight into the interactions of all, thus providing greater assurance that the complexities of the whole are appropriately considered and interrelated. Within this context, the committee views the Navy as an overall system, with fleet operations as the primary customer and shore installations and systems commands as major suppliers. Firmly establishing the *real* requirements of the fleet up front (i.e., what the customer is willing to pay for in the constrained budget environment) in any discussions of change will have a profound impact on what the shore installations and systems commands should be providing. The overall goal is to optimize the system.

Guide 6—Understand the Impact of Change on the People in the Organization

Change impacts people in an organization in the following ways: possible loss of self-image, disruption of emotional involvement and role relationships, and negative impact on the employee's sense of importance. Even an employee or person who suggests the idea for change recognizes there may be a negative impact on the role that he or she plays in the future system.

Psychologists have studied the stressful effects of change on people in a wide variety of circumstances. They define a stress curve that people go through when change in their "normal" activities occurs. The progressive stress stages that must be recognized and managed by leaders include (1) initial turmoil, (2) recoil against the change, (3) depression, (4) mourning about the change, (5) reacting to the change, and (6) reconstitution of the person's mental state to the new conditions. The degree to which each element of the stress curve is managed will determine the depth and duration of the stress curve. Even those who want and suggest change go through this cycle, although they usually experience a lesser degree of and time in depression, for example. These guides to successful change provide a framework for minimizing the effects of each part of the stress curve. The guides focus on generating early buy-in to and acceptance of change rather than on overcoming resistance to change, an activity that has a negative connotation, takes much longer and has higher costs, and too often leaves a bitter residue in the minds of the people involved.

The conventional approach to the development of a change tends to foster defensiveness (turmoil, recoil against change) rather than the openness and less stressful conditions needed. For example, starting a change effort by asking questions about what is wrong, whose fault it is, what measurements can be obtained to "show how bad things are," and who should be blamed for the poor performance creates a negative attitude almost immediately. A far more positive reaction can be obtained with different questions. Consider, as an illustration, the reaction of people to questions such as, How would you do it if you started all over again? The questions raised by these guides provide a way to get people to be comfortable with a situation so that change can proceed effectively. They also encourage a culture of continuous change because they show that there is never only one answer to the way we do things.

Change is always about people. Leveraging their core competencies in the organization is the crucial element in successful change. People are the critical factor in any specific system or subsystem (i.e., a ship, a submarine, or a weapon system) for it to operate effectively. Therefore, skilled sailors should spend a maximum of their time assigned to systems for which they were highly trained. The current practice of assigning new but highly trained sailors to mess duty aboard ship for extended periods creates a great deal of stress and adversely affects performance and reenlistment. Rather, a system perspective would strongly suggest that the galley (food delivery system) should be redesigned to require fewer people in order to deal with the shortage of younger personnel aboard ships.

Guide 7—Involve All the Stakeholders

This principle provides a basis for people with different interests and backgrounds who have a stake in the outcome to come together, work together, and change together. Involving all the stakeholders will provide a positive probing framework with different perspectives. This guide honors people and their ideas.

A comprehensive process or methodology for engaging all of the stakeholders in the creation of a shared vision and in the implementation of an action plan is required to accomplish enduring change in any organization. A successful process is one in which the participants contribute to and buy into the products and actions created. The process itself must include a proper sequence of meetings, topics, and events that have been shown to produce results. The critical elements of one such successful process are described in the following sections.

To have the maximum impact, the change process to be described should be initiated at the most senior leadership level of the organization and then repeated down through successive levels of leadership until the entire organization has been engaged. In the case of the Navy, the process should be initiated at the CNO level and should involve the fleet commanders, systems commands, and CNO staff. After this, fleet commanders should conduct similar processes with their fleet operators, the regional base commanders, the next level of systems commanders, and any other stakeholders critical to the organization. Following this, each command element, (e.g., each regional base commander) should conduct the process for his or her organization, carrying forward the work previously accomplished. Continuing this process down to the lowest level is the most efficient and lasting way to transform the overall organization.

The first step in the change process is to conduct a conference involving all of the stakeholders for the purpose of planning for the subsequent meetings. This is often referred to as the design conference. The objective is to identify critical issues, to articulate what the key focus is, and to design a process that will deal with the focus issue.

The next step in the process is for the stakeholders to create the vision of the future state discussed in detail in Guide 4. Following this would be the development of a set of strategic proposals or goals. These are creative proposals to move the organization in the new direction. The time for completing these major proposals may be many years, but significant segments with a scale of a few years should be identified. Within this shorter time frame, milestones can be set, measurements of progress can be made, and successes can be celebrated.

For example, a strategic proposal might be to design all future ships with the minimum number of personnel on board. Sailors drive many of the support requirements aboard ship and heavily impact the size and cost of shore installations. This strategic proposal would have a major cost reduction impact on future operations. This proposal would trigger one or more tactical proposals for accomplishing the goal such as increased use of automation, modification of on-board processes, and so on.

The stakeholders must also deal with the issues, blocks, deterrents, irritants, and so forth, that are obstacles in the path of accomplishing the vision. These contradictions may be structures, procedures, or attitudes that when clearly stated indicate the steps to be taken for their resolution. The team must deal with the root causes of these contradictions, not the superficial problems that often receive the greatest attention. The lack of money, personnel, or other resources is never the root obstacle impeding the path to the vision.

Many of the strategic proposals will address elimination of the obstacles lying in the path toward the vision. For example, although the Navy would like to change many processes, it currently cannot identify an accurate cost of these processes because the present accounting system does not provide sufficient details. A strategic proposal might be to obtain a new Navy-wide cost accounting system that would provide sufficient details to evaluate the cost of alternative processes.

The strategic proposals will be too numerous for any organization to tackle simultaneously. A prioritization must therefore be made. The highest-priority proposals should be implemented first. Task forces involving a cross section of the organization are then formed and charged with the responsibility for developing charters and tactics to actualize the selected strategic proposals on behalf of the entire organization.

The task forces should be charged with developing implementation calendars that identify the issues of where, when, and how key tactics will be implemented and by whom. The calendar should operate on a 90-day cycle, with major milestones identified at frequent intervals. The leadership team should review and approve the task force calendars.

Progress reviews should be conducted every 30 days to keep pressure on producing results and to ensure that all members are contributing. Peer pressure during these reviews can be a powerful motivator. If a task force successfully completes its objectives in the 90-day period to the satisfaction of the leadership team, victory is declared and a celebration should ensue. The task force would then create a new tactical proposal to address one of the strategic issues identified earlier, and the cycle would be repeated. If the proposal is not completed, an extension may be granted or a new task force assigned.

Progress toward the future state is accomplished in continuous improvement steps. Key quantitative metrics should be derived that will indicate the rate of progress. The quantitative goals to be accomplished over a period of time (e.g., annually) should be published and made highly visible to all members of the organization. Actual accomplishments as measured against the goals should be reported at frequent intervals so that the entire organization can judge progress.

The leadership must demand results in a timely manner. Groups and individuals must be held accountable for moving the organization forward toward the future state. Rewards for accomplishing goals and completing tasks should be used widely and frequently to create positive motivation for change. Experience has shown that large rewards are not necessary to motivate people. Small rewards and especially recognition carry great impact. Negative motivators in the form of punishment for failure to participate or cooperate are necessary but should be used sparingly and judiciously.

The teams and the organization should celebrate all significant victories in their path toward the future state. Celebrations build confidence, a sense of accomplishment, and motivation for tackling and succeeding in subsequent proposals.

The involvement of all stakeholders requires, in addition to the champions of change, facilitators or change agents using these guides to set up a basis for productive results from groups of people. However significant the champions of change may be, such skilled facilitators or change agents are necessary to obtain innovative results as well as successful change among all the smaller groups, task forces, and teams that the plan of action for change would set up.

Guide 8—Collect Only Essential Information

System performance measures are essential in the operation, management, and change of all organizations, but the term "essential" requires clarification. Suggesting, for example, that complete information collection about and measurements of existing conditions must start any change effort is a conventional approach that almost always exceeds what is essential. In addition, it leads to people's defensiveness, caused by the probing of analysts. Trying to collect all the data is impossible. Asking the stakeholders and related people about the purpose of any information or measurement collection and then placing that purpose into a context of larger ends should help determine if the data are worth collecting. Collecting information and obtaining measurements are not cost-free. Following this guide will limit the amount to be collected and will make what is collected much more relevant.

Guide 9—Recognize That Change Is Never Finished

Continuous improvement and betterment are necessary in making progress toward the target, maintaining the target when achieved, and using the target as the benchmark for future improvements. This guide even includes the scheduling of time to revise the future solution target. A critical aspect in this guide is to develop a change-resilient work force. People should look forward to change without the pressures of external forces. Any concept of having the "best" solutions now is the enemy of "better." Any change now is really a choice of what is considered the better alternative from among the options. Recognizing that change is never finished encourages the need to rethink day-to-day activities and to challenge (with these guides, not in a confrontational way) conventional wisdom about the purposes of doing things and the way needed things are done now.

Guide 10—Persevere in Seeking Change; It Takes a Long Time

The necessary cultural change in the Navy can take 5 to 7 years or longer. Installation of smaller system changes may also take many months to 1 to 2 years. Persevere, persevere, persevere should be the motto. It is critical to seek continuous, evolutionary—rather than only revolutionary—changes.

"Communicate" is a secondary mantra in this guide. Upward communication of needs and issues to higher levels of the Navy is often as important as the constant motivational and substantive messages to those below.

E

Committee Biographies

John D. Christie (Co-Chair) is a senior fellow at the Logistics Management Institute. Dr. Christie has an extensive background in DOD acquisition policy and program analysis. From 1989 to 1993, Dr. Christie was director of Acquisition Policy and Program Integration for the Under Secretary of Defense (Acquisition). In his role as director, Dr. Christie directed the preparation of a comprehensive revision of all defense acquisition policies and procedures resulting in the cancellation and consolidation of 500 prior separate issuances. He also prepared comprehensive acquisition program alternatives for the Secretary of Defense that resulted in multibillion-dollar budget reductions. As a former member of the Army Science Board, Dr. Christie was called upon to direct reviews of the Army analytical community and operations research activities for the Vice Chief of Staff, including the support of the overall Army acquisition process and its integration with the programming and budgeting process. Recently, Dr. Christie coauthored a paper on enhancing the success of DOD's outsourcing and led a team for the Commission on Roles and Missions of the Armed Forces that provided recommendations to improve defense management. Dr. Christie has also been an active participant in NRC studies; most recently, he served on the Panel on Statistical Methods for Testing and Evaluating Defense Systems.

John F. Egan (Co-Chair) retired this past spring as vice president for corporate development at Lockheed Martin Corporation. In recent years, Dr. Egan has been responsible for providing support to three successive chief executives in defining and implementing strategic plans to consolidate the defense industry. These include the merger of the Lockheed Corporation with the Martin Marietta Corporation and the acquisition of the defense segment of the Loral Corporation.

During these mergers, Dr. Egan provided leadership during the entire transaction cycle involving industry and market analysis, deal negotiations, antitrust filings, and transition planning and execution. An electrical engineer by training, Dr. Egan is a former chief scientist for the CNO and has extensive experience associated with the development of information technology. Dr. Egan currently is a member of the Chief of Naval Operations Executive Panel and the Naval Studies Board of the NRC.

John W. Asher III is chairman of Strategic Marketing and Analysis, Inc., and cofounder and chairman emeritus of Global Associates, Limited. Global Associates is a fast-growing small business specializing in advanced electronic products for both the public and the private sectors. At Global Associates, Mr. Asher has provided vision and strategic planning to grow a start-up company that was selected in 1997 as one of 50 FAST TRACK companies by *The Washington Business Journal*, Arthur Andersen, and Riggs Bank (selected 19 of 800). Besides having an extensive background in strategic planning and DOD acquisition, Mr. Asher has developed an enterprise information management system that has been successfully implemented in several large organizations. Mr. Asher also is a former naval officer, having retired at the rank of captain. His naval career included staff assignments to the deputy chief of Naval Operations for Submarines and as director of Submarine Operations and Tactical Readiness. Mr. Asher was also a senior member of the Submarine Research and Development Advisory Group assisting in the formulation of the Navy's future plans for weapons, sensors, and platforms.

Albert J. Baciocco, Jr., retired from the U.S. Navy as a vice admiral in 1987 after 34 years of distinguished service, principally within the nuclear submarine force and directing the Department of the Navy research and technology development enterprise. He graduated from the U.S. Naval Academy in 1953, receiving a bachelor of science degree in engineering. On retirement from the Navy, Admiral Baciocco formed the Baciocco Group, Inc., a technical and management consulting practice providing services to industry, primarily in areas of strategic planning, technology investment and application, and business planning and development. Admiral Baciocco is a director of Honeywell, Inc., and American Superconductor Corporation. In addition, he serves on several boards and committees of government, industry, and academe. He is a member of the Army Science Board of the Department of the Army, and of the NRC's Naval Studies Board. He serves on the boards of trustees of the University of South Carolina Research Institute and of the South Carolina Research Authority, a state-chartered, not-for-profit, advanced technology and management enterprise. In addition, he serves on the boards of directors of the Foundation for Research Development at the Medical University of South Carolina and of the Waste Policy Institute, an affiliate of Virginia Polytechnic Institute and State University, and currently serves as chair of the Southeastern Universities Research Association's Maritime Technical Advisory Committee to the Thomas Jefferson National Accelerator Laboratory in Newport News, Virginia.

Lloyd A. Duscha is an engineering consultant to various national and foreign government agencies, the World Bank, and private-sector clients. Mr. Duscha retired as deputy director, Engineering and Construction Directorate, of the U.S. Army Corps of Engineers, where he was responsible for the planning, engineering, and construction of all types of water resource projects and military facilities. His expertise includes project development and execution, dams, hydropower and hydraulic structures, building and underground technology, and coastal engineering and navigation. Mr. Duscha is a member of the National Academy of Engineering. He has served on the Board of Infrastructure and the Constructed Environment and has been an active participant in numerous NRC studies, including those focused on civil works research and development and technology and innovation in building. Most recently, he served on the Committee on Outsourcing of the Management of Planning, Design and Construction Services for Federal Facilities.

Elvin R. Heiberg III is president of Heiberg Associates, Inc. LG Heiberg (USA, Ret.) has an extensive background in facility construction and management. A retired chief of engineers for the Army (and commander of the Army Corps of Engineers), General Heiberg is well acquainted with engineering and construction issues, in both the public and private sectors, that relate to environmental engineering, infrastructure issues, privatization issues, and streamlined government acquisition of engineering and construction services. He is a member of the National Academy of Engineering. His NRC services include 2 years on the Transportation Research Board's executive committee, 6 years on the Commission on Engineering and Technical Systems, 4 years on the Board of Infrastructure and the Constructed Environment, and membership on the Board of Army Science and Technology in 1997. General Heiberg chaired the NRC's Federal Facilities Council until January 1998. He is a registered professional engineer (civil).

Samuel D. Kleinman is director of the Infrastructure and Readiness Team at the Center for Naval Analyses (CNA). Dr. Kleinman has an extensive background in the evaluation of business practices and, at CNA, is responsible for infrastructure and financial issues central to the Department of the Navy. His research interests include reducing the infrastructure, base consolidation, outsourcing, housing, management efficiency, industrial base, hollow force, acquisition reform, material support, and transportation. Dr. Kleinman has been a participant on several outside study teams. Some recent examples include the Defense Science Board's 1996 study of operating and support costs and the Navy's study on its acquisition organization.

Genie McBurnett retired from the Department of the Navy in 1996 as a member of the Senior Executive Service. Mrs. McBurnett's last assignment was as the assistant deputy chief of staff for Installations and Logistics, U.S. Marine

Corps. In this position, Mrs. McBurnett was responsible for policy formulation and oversight of Marine Corps logistics, facilities, environmental management, and base closure. Her previous positions include deputy commander, Space and Naval Warfare Systems Command, and principal deputy assistant secretary of the Navy (research, development, and acquisition). She also served as a member of the Navy's Base Structure and Evaluation Committee responsible for review of all Department of the Navy infrastructure and recommendations for base realignment and closure.

Nancy Y. Moore is a senior engineer at RAND, where her research interests include water policy, military logistics, and business practices. Dr. Moore has conducted studies on efficient ground and surface water use in California, sustainability of the U.S. Air Force post-attack command and control aircraft, and improving materiel distribution for the Army. She recently led a Defense Logistics Agency study on strategic approaches to outsourcing. She is currently examining outsourcing and contracting issues for the Air Force. A water resources engineer by training, Dr. Moore has also been involved in studying alternative water resources policies, particularly water marketing. Her participation in NRC studies has included membership on the Committee on Flood Control Alternatives in the American River Basin and the Glen Canyon Environmental Studies Review.

Gerald Nadler is professor emeritus of industrial and systems engineering and IBM chair emeritus in engineering management at the University of Southern California. Dr. Nadler has an extensive background in human resources, where his major interest is the study of methodologies and reasoning processes used by leading engineers, planners, designers, and other professionals who create new, and structure existing, systems and solutions (e.g., products, operating procedures, organizational arrangements, information flows, and facilities flow). Dr. Nadler is a member of the National Academy of Engineering and has served on NRC committees, including the Committee on Human Resources, Organizations, and the Adoption of Workplace Technologies.

Sean O'Keefe is the Louis A. Bantle Professor of Business and Government Policy in the Maxwell School of Citizenship and Public Affairs at Syracuse University. Professor O'Keefe is also director of national securities studies—a joint partnership between Syracuse University and the Johns Hopkins University for the delivery of senior executive education programs and case study development for the Department of Defense. Professor O'Keefe's background is primarily related to national security policy and financial management. He was appointed Secretary of the Navy in 1992 and has served as comptroller and chief financial officer of DOD since 1989 as a member of Secretary Dick Cheney's Pentagon management team. Previously, he served on the U.S. Senate Committee on Appropriations staff as staff director of the Defense Subcommittee. He is a fellow of the National Academy of Public Administration and a member of several corporate and advisory boards.

Dan R. Olsen, Jr., is director of the Human Computer Interaction Institute and professor of computer science at Carnegie Mellon University. Dr. Olsen has an extensive background in state-of-the-art computer technologies. At Carnegie Mellon, Dr. Olsen's principal areas of research and teaching include automatic generation of graphical user interfaces, high-level programming languages, man-machine communication, network-based interaction, and visual and graphical programming systems. Previously, Dr. Olsen was chairman of the Department of Computer Science at Brigham Young University. His professional activities include vice chair of publications for the Association for Computing Machinery's Special Interest Group on Computer-Human Interaction and associate editor of *Human-Computer Interaction and Interactions.*

Michael W. O'Neill is a national director of collaborative operations resources at Deloitte and Touche, LLP. A certified public accountant, his 25 years' experience include management consulting, outsourcing, daily operations, and the management of large, complex projects involving financial operations and data processing. At Deloitte and Touche, Mr. O'Neill is responsible for all accounting and outsourcing projects for the federal government as they relate to financial operations. In addition, he is the partner in charge of a recently awarded engagement with the Government National Mortgage Association to perform compliance reviews and agreed-upon review. Prior to joining Deloitte and Touche, Mr. O'Neill served as a worldwide partner and a member of the board of directors of William M. Mercer, Inc. In this role, he was responsible for all administrative consulting, outsourcing operations and technology applications, and business units of the firm.

Herbert Rabin is director of the Engineering Research Center, professor of electrical engineering, and associate dean of the A. James Clark School of Engineering at the University of Maryland, College Park. Dr. Rabin's professional experience has focused on research and applied technology in the fields of solid-state physics, nonlinear optics, and space research and development. Dr. Rabin has held a number of positions at the Naval Research Laboratory, including associate director for Space and Communications Science and Technology. In addition, he served as deputy assistant secretary of the Navy for Applied Research and Space Technology. Dr. Rabin is a member of the NRC's Naval Studies Board, and a fellow of the American Physical Society, Optical Society of America, American Association for the Advancement of Science, and American Institute of Aeronautics and Astronautics.

Joseph B. Reagan is retired vice president and general manager of research and development at Lockheed Martin Missile and Space Co. and was a corporate officer of the Lockheed Martin Corporation. As general manager, he led more than 750 scientists and engineers in the development of advanced technologies in the fields of optics, electrooptics, software, guidance and controls, electronics, and materials. He also led a major cultural change process within research and development that transformed the organization into a more successful, business-oriented entity. Dr. Reagan is currently a director of Southwall Technologies, Inc., a high-technology company specializing in the manufacturing of thin-film coatings for high-performance residential, industrial, and automotive windows. He is also a director on the board of the Tech Museum of Innovation, where he is the chairman of the Education Committee. Dr. Reagan is also a senior adviser to Knowledgen, Inc., a startup company that provides software products to facilitate the conversion of information into knowledge. He is involved in numerous activities that foster the improvement of science and mathematics education in the United States. Dr. Reagan is a member of the National Academy of Engineering, a fellow of the American Institute of Aeronautics and Astronautics, and a member of the NRC's Naval Studies Board.

John M. Stewart is director of McKinsey and Company, Inc. Mr. Stewart has been employed at McKinsey for more than 35 years, where he has specialized in business administration and management consulting. His responsibilities today include advising top management of leading organizations on issues of strategy, organization, and operations. In addition, he is actively involved in the consulting of international business competition, research and development, and factory operations. Prior to joining McKinsey, Mr. Stewart was a program manager in aerospace at TRW. Mr. Stewart has served with a number of boards, including the Defense Science Board and the Manufacturing Science Board of the National Research Council. He recently served in the Defense Science Board's Task Force on Outsourcing and Privatization.

Raymond M. Walsh, a retired rear admiral USN, is vice president at Sonalysts, Inc. Since joining Sonalysts, Admiral Walsh has been a lead analyst and project leader for several forward-looking studies. Among other projects, he is currently assisting in the development of the command and control design for the AEGIS weapons system for Tactical Ballistic Missile Defense. Admiral Walsh's broad range of experience includes the command of two surface combatants as a naval surface warfare officer and as an operations analyst ashore involved with Department of the Navy budgeting, planning, and programming processes. Admiral Walsh was also director of the Operations Division for the Office of Budget and Reports under the Navy comptroller, where he was the responsible official for all Department of the Navy operating budget accounts.

Mitzi M. Wertheim is a consultant for enterprise solutions at the Center for Naval Analyses (CNA). Before joining the CNA, she was vice president of enterprise solutions at SRA International, Inc. Ms. Wertheim's expertise is in recognizing reengineering needs, creatively applying business process reengineering methods and tools, and teaching large corporations to increase service while reducing cost. At SRA, Ms. Wertheim creatively applied business process reengineering methods to improve productivity and provide higher levels of service, while reducing costs for large corporations to become customer focused, process focused, and team oriented using information technology as an enabler. Her responsibilities included identifying linkages and interdependencies in organizations and then leveraging information technology to achieve business objectives. Prior to joining SRA, Ms. Wertheim worked with IBM Federal Systems Company as an enterprise consultant, marketing manager, program manager, and technical assistant. From 1977 to 1981, Ms. Wertheim was the deputy undersecretary of the Navy. Ms. Wertheim is involved with a number of outside organizations, including the Council of Foreign Relations and the Advisory Board of the Defense Budget Group. She is the founder of the MIT Seminar XXI, Foreign Policy and National Security, now in its 13th year. She is a member of the NRC's Naval Studies Board.

F

Meetings Agendas

FEBRUARY 4-5, 1998
WASHINGTON, D.C.

Wednesday, February 4, 1998

CLOSED SESSION (COMMITTEE MEMBERS AND NRC STAFF ONLY)

8:30 a.m. CONVENE—Welcome, Administrative Issues, Composition and Balance Discussion, Overview
Dr. John D. Christie, Study Co-Chair
Dr. John F. Egan, Study Co-Chair
Dr. Ronald D. Taylor, NSB Director

SHORE INSTALLATION INFRASTRUCTURE

10:30 SHORE SUPPORT INFRASTRUCTURE—Areas of Interest and Strategic Vision
VADM William J. Hancock, USN
Deputy Chief of Naval Operations, Logistics, N4

11:15 REINVENTING SHORE INFRASTRUCTURE
RADM John T. Scudi, USN
Director, Shore Installation Management Division, N46

12:00 p.m. INFRASTRUCTURE COST REDUCTION INITIATIVE—Regionalization, Installation Claimant Consolidation, Infrastructure Vision and Strategic Plan **(Working Lunch)**
Mr. David M. Wennergren
Head, Plans and Policy Branch (Regionalization and Infrastructure Reduction), N464

NAVY INITIATIVES

1:15 REGIONAL MAINTENANCE—Core Capability, Infrastructure, Comply with Aircraft/Ship Maintenance Plans, Flight/Ship Safety, Technical Authority, Employ Business Cost Analyses for Commodity Reviews
CAPT Sharon Gurke, USN
Head, Industrial Facilities, Policy and Resource Branch, N431

2:15 SMART BASE—Overview, Smart Link, Distance Learning, Performance Based Management Support System, Electronic Commerce/Electronic Data Interchange, Security, Smart Card
CAPT Dennis Parsons, USN, Team Leader, Smart Base Project, NSWCCD, N466
CDR Mike Butler, USN, Team Member, Smart Base Project, NSWCCD, N466C

FUTURE STUDY PLANS AND ISSUES TO BE ADDRESSED

4:00 COMMITTEE DISCUSSION

3:00 END SESSION

Thursday, February 5, 1998

8:30 a.m. CONVENE—Discussion and Review
Dr. John D. Christie, Study Co-Chair
Dr. John F. Egan, Study Co-Chair

SHORE INSTALLATION MANAGEMENT

9:00 LOGISTICS INITIATIVES
CDR Jerry Francom, N412H

9:45 FACILITY MANAGEMENT—Real Property Maintenance, Housing, Public Private Ventures, Demolition
CDR Mike Blount, USN, Military Construction Branch, N445C
CDR Tom Hollinberger, USN, Housing/BQ Branch, N443
CDR Rick Roth, USN, Head, Real Property Maintenance and Energy Branch, N442

BUSINESS PRACTICES

11:00 OUTSOURCING—Reduction of Resources, Industry Form, Redefine Management Information Database, Develop FY98 Congressional Announcement

CAPT Hugh McCullom, USN
Deputy, Outsourcing Division, N47B

1:00 p.m. PRIVATIZATION—Tool Box

CDR Steven Muck, USN
Plans and Policy Branch (Regionalization and Infrastructure Reduction), N464E

2:00 CREATING A REVOLUTION IN BUSINESS AFFAIRS IN DOD

Dr. Sam Kleinman
Director, Infrastructure and Readiness Team, Center for Naval Analyses

FUTURE STUDY PLANS AND ISSUES TO BE ADDRESSED

3:00 COMMITTEE DISCUSSION

5:00 ADJOURN

MARCH 4-5, 1998
WASHINGTON, D.C.

Wednesday, March 4, 1998

8:30 a.m. CONVENE—Welcome, Administrative Issues, Discussion, Overview

Dr. John D. Christie, Study Co-Chair
Dr. John F. Egan, Study Co-Chair

MANAGERIAL ACCOUNTING

9:00 ACTIVITY BASE COSTING—Naval Air Force Pacific Command, Drill Down Demonstration

CAPT Thomas Conroy, Jr., USN, Material Officer, NAVAIRPAC
CDR Martin J. Shaunessy, USN, Flying Hours Program Manager, NAVAIRPAC

11:00 PERFORMANCE-BASED MANAGEMENT FOR INSTALLATIONS—Shore Installation Initiatives

CDR Nicholas W. Zimmon, USN, PBMMS, Smart Base Project, N462B

12:00 p.m. HEWLETT-PACKARD COMPANY DEMONSTRATION—Cost Tracking

Mr. Michael A. Johnson, Facility Manager, Hewlett- Packard Company

NAVY PLANNING, PROGRAMMING, AND BUDGETING INITIATIVES

1:00 PLANNING, PROGRAMMING, AND BUDGET INITIATIVES—OPNAV(N8)
Organizational Overview, Financial Processes, and Initiatives
RADM Jay B. Yakeley III, USN, N80, Director, Programming Division

2:45 DEPARTMENT OF THE NAVY SUPPORT AND INFRASTRUCTURE
CAPT David Dealy, USN, CNO N4J, Program and Budgeting

4:00 FUND FLOW AND REVISED INFRASTRUCTURE BREAKOUT —SECDEF Guidance, Functions, Appropriations, and Claimants
RADM Raymond M. Walsh, USN (Ret.), Vice President, Sonalysts, Inc.

FUTURE STUDY PLANS AND ISSUES TO BE ADDRESSED

5:00 COMMITTEE DISCUSSION

5:15 RECEPTION AND DINNER (South Prefunction Area and GR 118)
VADM Stephen F. Loftus, USN (Ret.), Guest Speaker

7:00 END SESSION

Thursday, March 5, 1998

8:30 a.m. CONVENE—Discussion and Review
Dr. John D. Christie, Study Co-Chair
Dr. John F. Egan, Study Co-Chair

FACILITY MANAGEMENT

9:00 FACILITY MANAGEMENT OUTSOURCING—Buyer/Provider Services in the Commercial Sector
Mr. Paul A. Uber, Senior Vice President, Compass Management and Leasing
Mr. H. Inge Waddle, Assistant Vice President, Bell South Telecommunications

11:30 COMMITTEE DISCUSSION

INFORMATION TECHNOLOGY

1:15 p.m. THE FUTURE OF INFORMATION TECHNOLOGY IN THE USN—Navy Infrastructure Initiatives
RADM John A. Gauss, USN, Prospective Commander, Naval Space System Warfare

3:00 SMART LINK
Mr. Robert M. Johnson, Technical Director, Smart Link/ Smart Base Project, N466TD

FUTURE STUDY PLANS AND ISSUES TO BE ADDRESSED

4:00 COMMITTEE DISCUSSION

5:00 ADJOURN

APRIL 27-30, 1998
SAN DIEGO, CALIFORNIA

Monday, April 27, 1998

8:15 a.m. CONVENE—Welcome, Discussion, Introductions
Dr. John D. Christie, Study Co-Chair
Dr. John F. Egan, Study Co-Chair

CHANGE MANAGEMENT

8:30 IMPLEMENTING CHANGE WITHIN AN ORGANIZATION—Creative Thinking and Innovative Action
Mr. Lee Early, President, LE Associates

10:15 THE SYMBOLIC MEANING OF CHANGE
Dr. Harry Levinson, Chairman, The Levinson Institute

11:45 **Working Lunch**
Committee Discussion with San Diego Chamber of Commerce Members

FAMILY HOUSING

1:45 p.m. FAMILY HOUSING PRIVATIZATION CONCEPT FOR SAN DIEGO REGION
CAPT Jill Usher, USN, Assistant Chief of Staff, Housing

CHANGE AGENTS

3:30 ACTIVITY BASE COSTING
Mr. George F. Murphy, Senior Manager, KPMG Peat Marwick LLP

DISCUSSION AND SITE TOUR

4:30 DISCUSSION

5:00 BUS DEPARTS FOR SHIP TOUR AT SAN DIEGO NAVAL STATION

6:30 END SESSION

Tuesday, April 28, 1998

8:15 a.m. CONVENE—Welcome, Discussion, Review, Introductions
Dr. John D. Christie, Study Co-Chair
Dr. John F. Egan, Study Co-Chair

USN SAN DIEGO REGIONALIZATION—PLAN FOR RESTRUCTURING

8:30 a.m. REGIONAL COMMAND OVERVIEW—Plan for Restructuring, Regional Implementation, Representative Command Masterchief Perspective
RADM Veronica Z. Froman, USN, Commander, Naval Base San Diego
Business Office Group
Representative Command Master Chief Perspective

USN SAN DIEGO REGIONALIZATION—READINESS AND MANAGEMENT

11:05 LEADING CHANGE IN THE SOUTHWEST
RADM Veronica Z. Froman, USN, Commander, Naval Base San Diego

11:55 **Working Lunch**
CO'S ROLE PERSPECTIVE—Naval Base San Diego and Waterfront Operations
CAPT Bryan L. Rollins, USN, Chief of Staff, Commander, Naval Base San Diego
CAPT Vincent E. Smith, USN, Commanding Officer, Naval Station San Diego
CDR Ed Caviness, USN, Waterfront Operations Officer, Naval Station San Diego

1:30 p.m. NAVAL BASE SAN DIEGO USER PERSPECTIVE—Regional Freight Transportation
LCDR Pil Allison, USN, Fleet and Industrial Supply Center San Diego

2:15 REGIONAL ADVISORY BOARD ROUNDTABLE DISCUSSION
Regional Installation Commanders and Primary Service Providers Cos
Business Office Brief on Current/Future Management Structure and Relationships

SITE TOUR

4:30 BUS DEPARTS FOR SUBMARINE TOUR

6:00 END SESSION

Wednesday, April 29, 1998

8:15 a.m. CONVENE—Welcome, Discussion, Review, Introductions
Dr. John D. Christie, Study Co-Chair
Dr. John F. Egan, Study Co-Chair

USN SAN DIEGO REGIONALIZATION—READINESS AND MANAGEMENT

8:30 INFORMATION TECHNOLOGY—San Diego Regional Initiatives
CAPT Judith A. Tarbox, USN, Commanding Officer, Naval Computer and Telecommunication Station, San Diego

9:45 NAVAL BASE SAN DIEGO USER PERSPECTIVE—Southwest Regional Maintenance Center
CDR B. Hernandez, USN, SWRMC Maintenance Analysis Group

USN SAN DIEGO REGIONALIZATION—FACILITIES MANAGEMENT

11:15 NAVAL BASE SAN DIEGO FUNCTION PERSPECTIVE—Service to Region (Program Manager of Public Works)
CAPT A.D. Brunhart, USN, Commander, Navy Public Works Center, SUBASE SD

12:15 p.m. **Working Lunch**
IMPLEMENTATION OF MAXIMO AND ABM IN FACILITIES MANAGEMENT
CDR Thomas Bersson, USN, CEC, Public Works Officer, Port Hueneme
ENS Holly Jenkins, USN, Public Works Officer, Port Hueneme

1:15 NAVAL BASE SAN DIEGO FUNCTION PERSPECTIVE—Service to Region (Planning Support Role to Facilities/Environmental)
CAPT T. Heinrichs, USN, CEC, Executive Officer, Naval Facilities Engineering Command, Southwest Division

CLOSED SESSION (COMMITTEE MEMBERS AND NRC STAFF ONLY)

2:30 CLOSED SESSION (COMMITTEE DISCUSSION AND NRC STAFF ONLY)

SITE TOUR

4:30 BUS DEPARTS FOR WINDSHIELD TOUR OF NADEP FACILITY, SQUADRON, AND USS *KITTY HAWK*

Thursday, April 30, 1998

8:15 a.m. CONVENE—Welcome, Discussion, Review, Introductions
Dr. John D. Christie, Study Co-Chair
Dr. John F. Egan, Study Co-Chair

USN SAN DIEGO REGIONALIZATION—OUTSIDE SERVICES

8:30 SERVICE PROVIDED BY PRIVATE CONTRACTOR TO NAVAL BASE SAN DIEGO—Service to Region
Ms. Sheila Wasson, Fleet and Industrial Supply Center SD
Mr. Michael Zeiders, President, Zeiders Enterprises Incorporated (FSC Contractor)
Mr. John Kays, Sales Manager, Corporate Express (FISC-SD Regional Supplies Contractor)

USN SAN DIEGO REGIONALIZATION—PLAN FOR RESTRUCTURING

10:15 COMMITTEE DISCUSSION WITH REGIONAL COMMANDER, CHIEF OF STAFF, AND REGIONAL BUSINESS MANAGER

RADM Veronica Z. Froman, USN, Commander, Naval Base San Diego

CAPT Bryan Rollins, USN, Chief of Staff, Naval Base San Diego

CAPT William E. Bickert, Jr., SC, USN, Business Manager, Naval Base San Diego

CLOSED SESSION (COMMITTEE MEMBERS AND NRC STAFF ONLY)

12:00 p.m. LUNCH (CONTINUED COMMITTEE DISCUSSION)

FUTURE STUDY PLANS AND ISSUES TO BE ADDRESSED

1:00 COMMITTEE DISCUSSION

5:00 ADJOURN

MAY 20-21, 1998
WASHINGTON, D.C.

Wednesday, May 20, 1998

8:30 a.m. CONVENE—Welcome, Administrative Issues, Discussion, Overview

Dr. John D. Christie, Study Co-Chair

Dr. John F. Egan, Study Co-Chair

NAVY REGIONALIZATION—HAMPTON ROADS REGION

8:45 REGIONALIZATION IN HAMPTON ROADS—Overview of Hampton Roads Region, Consolidation Efforts, Plan for Implementation

RADM R. Tim Zeimer, USN, Commander, Naval Base Norfolk

Mr. Thomas Crabtree, Business Management Officer, PWC Norfolk

NAVY MAINTENANCE AND SUPPLY—CONSOLIDATION AND COST REDUCTION INITIATIVES

12:30 p.m. **Lunch Roundtable Discussion**

NAVY MAINTENANCE—Consolidation and Cost Reduction Initiatives

RADM George R. Yount, USN, Director, Support, Maintenance Modernization (N43)

CAPT Sharon Gurke, USN, Head, Industrial Facilities, Policy, & Resource (N431)

2:00 BREAK

2:15 NAVY SUPPLY —Consolidation and Cost Reduction Initiatives
Mr. Larry Glasco, SES, Assistant Deputy Commander for Fleet Logistics, Naval Ship Systems Command

INFORMATION SYSTEMS AND MANAGEMENT ON BASES—USAF PERSPECTIVE

3:45 INFORMATION SYSTEMS AND MANAGEMENT
BrigGen (Sel) Bud Bell, USAF, Director of Systems, Communications and Information Center

5:30 END SESSION

Thursday, May 21, 1998

CLOSED SESSION (COMMITTEE MEMBERS AND NRC STAFF ONLY)

8:30 a.m. CONVENE—Discussion and Review
Dr. John D. Christie, Study Co-Chair
Dr. John F. Egan, Study Co-Chair

9:00 COMMITTEE DISCUSSION

1:00 p.m. FUTURE STUDY PLANS AND ISSUES TO BE ADDRESSED

5:00 ADJOURN

JUNE 10-11, 1998
WASHINGTON, D.C.

Wednesday, June 10, 1998

8:30 a.m. CONVENE—Welcome, Administrative Issues, Discussion, Overview
Dr. John D. Christie, Study Co-chair
Dr. John F. Egan, Study Co-chair

NAVAL EDUCATION AND TRAINING

8:45 NAVAL EDUCATION AND TRAINING
Dr. Allen Zeman, Deputy Director, CNET, N7B

FLEET READINESS

10:15 TYPE COMMANDER'S PERSPECTIVE
VADM Henry C. Giffin III, USN, Commander, Naval Surface Force, U.S. Atlantic Fleet

CHANGE MANAGEMENT

12:15 p.m. **Working Lunch**
IMPLEMENTING CHANGE WITHIN A CONSTRAINED ENVIRONMENT
Mr. Terry Baker, Director of Enterprise Development
Office of the Mayor, Indianapolis, Indiana

OPEN SESSION

2:00 COMMITTEE DISCUSSION—Debrief of NAVSTA Pascagoula Subcommittee Visit

CLOSED SESSION (COMMITTEE MEMBERS AND NRC STAFF ONLY)

3:00 COMMITTEE DISCUSSION

5:00 END SESSION

Thursday, June 11, 1998

8:15 a.m. CONVENE—Discussion and Review
Dr. John D. Christie, Study Co-Chair
Dr. John F. Egan, Study Co-Chair

INFORMATION TECHNOLOGY IN THE DEPARTMENT OF THE NAVY

8:30 INFORMATION TECHNOLOGY IN THE DEPARTMENT OF THE NAVY
Dr. Ann Miller, Chief Information Officer, Department of the Navy

NAVAL MANPOWER PLANNING

10:45 TOTAL FORCE PLANNING—Manpower Plans, Sea-Shore Rotation, Information Services
RADM (Sel) Peter W. Marzluff, USN, Director, Total Force Programming Manpower and Information Resource Management
CAPT (Sel) Dennis Haines, USN

INFORMATION TECHNOLOGY FOR THE 21ST CENTURY

12:15 p.m. **(Working Lunch)**
INFORMATION TECHNOLOGY FOR THE 21ST CENTURY
RADM Thomas J. Elliot, Jr., USN, Deputy Director, Submarine Warfare Division, N87B

OPEN SESSION

3:00 COMMITTEE DISCUSSION—Team Meetings

5:00 ADJOURN

JULY 1-2, 1998
WASHINGTON, D.C.

Wednesday, July 1, 1998

8:30 a.m. CONVENE—Welcome, Administrative Issues, Discussion, Overview
Dr. John D. Christie, Study Co-Chair
Dr. John F. Egan, Study Co-Chair

	CLOSED SESSION (COMMITTEE MEMBERS AND NRC STAFF ONLY)
9:00	COMMITTEE DISCUSSION
12:00 p.m.	LUNCH
1:00	COMMITTEE DISCUSSION
5:00	END SESSION

Thursday, July 2, 1998

8:30 a.m.	CONVENE—Welcome, Administrative Issues, Discussion, Overview Dr. John D. Christie, Study Co-Chair Dr. John F. Egan, Study Co-Chair
	CLOSED SESSION (COMMITTEE MEMBERS AND NRC STAFF ONLY)
9:00	COMMITTEE DISCUSSION
12:00 p.m.	LUNCH
1:00	COMMITTEE DISCUSSION
5:00	ADJOURN

G

Acronyms and Abbreviations

ABC	Activity-based costing
ABM	Activity-based management
ABS	Activity-based spending
ACOS	Assistant Chief of Staff
ASN/FM	Assistant Secretary of the Navy, Financial Management and Comptroller
ASN/I&E	Assistant Secretary of the Navy, Installation and Environment
ASN/RDA	Assistant Secretary of the Navy, Research, Development, and Acquisition
ATBM	Advanced tactical ballistic missile
ATM	Asynchronous transfer mode
BAM	Bandwidth Baseline Assessment Memorandum
BAS	Basic allowance for subsistence
BOS	Base operating support
BRAC	Base realignment and closure
CEC	Civil Engineer Corps
CINC	Commander in Chief
CINCLANT	Commander in Chief, Atlantic
CINCLANTFLT	Commander in Chief, Atlantic Fleet
CINCPACFLT	Commander in Chief, Pacific Fleet
CIO	Chief Information Officer
CNA	Center for Naval Analyses
CNET	Commander, Navy Education and Training
CNO	Chief of Naval Operations

CONUS	Continental United States
COTS	Commercial off-the-shelf
DBOF	Defense Business Operations Fund
DCNO	Deputy Chief of Naval Operations
DD	Destroyer
DDDC	Defense Distribution, San Diego, California
DOD	Department of Defense
ERP	Enterprise resource planning
FISC	Fleet Industrial Supply Center
FY	Fiscal year
FYDP	Future Years Defense Plan
GAO	General Accounting Office
HTML	Hypertext Mark-up Language
HTTP	Hypertext Transfer Protocol
IT	Information technology
IT-21	Information Technology for the 21st Century initiative
JCS	Joint Chiefs of Staff
JSF	Joint strike fighter
LAN	Local area network
LANTFLT	Atlantic Fleet
MAN	Metropolitan area network
MEO	Most efficient organization
MOU	Memorandum of Understanding
MPN	Military personnel
NADEP	Naval aviation depot
NAVAIR	Naval Air Systems Command
NAVFAC	Naval Facility Engineering Command
NAVSEA	Naval Sea Systems Command
NAVSUP	Naval Supply Systems Command
NAWC	Naval Air Warfare Center
NCBC	Naval Construction Battalion Center
NEC	Navy enlisted classification (code)
NOR	Net operating results
NVI	Naval Virtual Intranet
NWCF	Navy Working Capital Fund
O&M	Operations and maintenance
OMB	Office of Management and Budget
OPNAV	Office of the Chief of Naval Operations
OPNAV-N4	Office of the Chief of Naval Operations for Logistics
OSD	Office of the Secretary of Defense
PA&E	Program Analysis and Evaluation
PACFLT	Pacific fleet
PC	Personal computer

PDF	Portable document format
PSD	Personnel support detachment
psi	Pounds per square inch
PWC	Public Works Center
PWS	Performance work statement
QOL	Quality of life
RDT&E	Research, development, test, and evaluation
RPM	Real property maintenance
SECNAV	Secretary of the Navy
SIMA	Shore intermediate maintenance activity
SPAWAR	Space and Naval Warfare Systems Command
TIMS	Transportation Information Management System
TOA	Total obligational authority
USMC	U.S. Marine Corps
VCNO	Vice Chief of Naval Operations
WAN	Wide area network
WWW	World Wide Web
Y2K	Year 2000 computer issues